SpringerBriefs in Energy

SpringerBriefs in Energy presents concise summaries of cutting-edge research and practical applications in all aspects of Energy. Featuring compact volumes of 50 to 125 pages, the series covers a range of content from professional to academic. Typical topics might include:

- A snapshot of a hot or emerging topic
- A contextual literature review
- A timely report of state-of-the art analytical techniques
- An in-depth case study
- A presentation of core concepts that students must understand in order to make independent contributions.

Briefs allow authors to present their ideas and readers to absorb them with minimal time investment.

Briefs will be published as part of Springer's eBook collection, with millions of users worldwide. In addition, Briefs will be available for individual print and electronic purchase. Briefs are characterized by fast, global electronic dissemination, standard publishing contracts, easy-to-use manuscript preparation and formatting guidelines, and expedited production schedules. We aim for publication 8–12 weeks after acceptance.

Both solicited and unsolicited manuscripts are considered for publication in this series. Briefs can also arise from the scale up of a planned chapter. Instead of simply contributing to an edited volume, the author gets an authored book with the space necessary to provide more data, fundamentals and background on the subject, methodology, future outlook, etc.

SpringerBriefs in Energy contains a distinct subseries focusing on Energy Analysis and edited by Charles Hall, State University of New York. Books for this subseries will emphasize quantitative accounting of energy use and availability, including the potential and limitations of new technologies in terms of energy returned on energy invested. The second distinct subseries connected to SpringerBriefs in Energy, entitled Computational Modeling of Energy Systems, is edited by Thomas Nagel, and Haibing Shao, Helmholtz Centre for Environmental Research - UFZ, Leipzig, Germany. This sub-series publishes titles focusing on the role that computer-aided engineering (CAE) plays in advancing various engineering sectors, particularly in the context of transforming energy systems towards renewable sources, decentralized landscapes, and smart grids.

All Springer brief titles should undergo standard single-blind peer-review to ensure high scientific quality by at least two experts in the field.

Zhaohao Li · Jinjin Yang · Hongming Fu ·
Haiping Chen

Hydrophobic Ceramic Membranes for CO$_2$ Capture

Advancing Clean Energy Technologies

 Springer

Zhaohao Li
School of Energy, Power and Mechanical
Engineering
North China Electric Power University
Beijing, China

Hongming Fu
China Energy New Energy Technology
Research Institute Co., Ltd.
Beijing, China

Jinjin Yang
School of Energy, Power and Mechanical
Engineering
North China Electric Power University
Beijing, China

Haiping Chen
School of Energy, Power and Mechanical
Engineering
North China Electric Power University
Beijing, China

ISSN 2191-5520   ISSN 2191-5539 (electronic)
SpringerBriefs in Energy
ISBN 978-3-031-77677-9   ISBN 978-3-031-77678-6 (eBook)
https://doi.org/10.1007/978-3-031-77678-6

This work was supported by China Postdoctoral Science Foundation (Project No.: 2022M711133).

This Springer imprint is published by the registered company Springer Nature Switzerland AG
The registered company address is: Gewerbestrasse 11, 6330 Cham, Switzerland

If disposing of this product, please recycle the paper.

Preface

Aiming at the problem that membrane materials are easy to be wetted, a ceramic membrane with hydrophobic properties is prepared for CO_2 capture in membrane contactors, in order to provide a new idea for CO_2 capture technology in thermal power and other industries, and promote the clean and efficient development of thermal power industry. The preparation of ceramic membranes, the mechanism of mass transfer in porous media and the mechanism of CO_2 capture are studied in detail by means of experiment and theoretical analysis. The performance of commercial Al_2O_3 ceramic membrane and traditional polytetrafluoroethylene membrane for CO_2 capture is compared and analyzed, and the feasibility of ceramic membrane for CO_2 capture is verified. In order to solve the problem of increasing mass transfer resistance caused by membrane wetting, which leads to deterioration of membrane performance, and further improve the CO_2 capture efficiency of ceramic membrane, hydrophobic and super-hydrophobic ceramic membranes are developed respectively. The effects of different intake conditions and different absorbent flow rates on the CO_2 capture performance of ceramic membrane contactors are experimentally studied. The mass transfer mechanism and mass transfer resistance distribution in the process of CO_2 capture by ceramic membrane are investigated. Considering the application cost of commercial ceramic membrane, superhydrophobic ceramic membrane based on power plant waste fly ash is developed, and its CO_2 capturing performance and surface dynamic characteristics are discussed in detail.

Beijing, China
October 2024

Zhaohao Li
Jinjin Yang
Hongming Fu
Haiping Chen

Acknowledgement This book was financially supported by the China Postdoctoral Science Foundation (Project No.: 2022M711133).

Contents

About the Authors

Zhaohao Li received his doctor's degree from North China Electric Power University in June 2021. He focuses his research interests on the CO_2 capture technology. He has published 11 journal papers in these fields as the first or the corresponding author. Furthermore, he has published 1 SpringerBriefs in Energy.

Jinjin Yang is currently an undergraduate at the School of Energy, Power and Mechanical Engineering in North China Electric Power University. She focuses on the membrane absorption technology.

Hongming Fu received his doctor's degree from North China Electric Power University in June 2024. He currently works in the China Energy New Energy Technology Research Institute Co., Ltd. as a researcher, and mainly focuses his research interests on the preparation and characterization of ceramic membranes. He has published 5 journal papers in these fields as the first author.

Haiping Chen is a professor of North China Electric Power University, doctoral supervisor, Deputy Director of the National Thermal Power Engineering Technology Research Center, project leader of the National key research and development plan, and energy conservation expert of the National Energy Conservation Center and China Electricity Union. He has been engaged in research on modern energy-saving theory and power plant pollutant emission reduction technology for a long time, and presided over the completion of more than 50 projects such as the national key research and development plan project "Research and application of key technologies for water efficient and low-cost recovery and treatment of coal-fired generating units", and the strategic research and consulting project of the Chinese Academy of Engineering "Construction of modern energy system evaluation system and quantitative comparative study by province". The relevant achievements have won the first prize of China Electric Power Science and Technology Progress, the second prize of Science and Technology Progress of the Ministry of Education, the third prize of science and Technology Progress of Hebei Province, and the first prize of China Electric Power Innovation Award. Related achievements have published more than 100 academic papers, more than 10 authorized patents, and participated in the preparation of 3 industry standards.

Chapter 1
Introduction

This chapter will show the background of the application of capturing CO_2 by the ceramic composite membranes, including the current development status and future development trends. Then, on this basis, the study of membrane preparation and mass transfer mechanism will be reviewed, and the outstanding problems will be summarized. Finally, the structure of the book will be described.

1.1 Background

Energy is an important cornerstone of the human social and the economic development. Fossil fuels play the decisive role in the development of the world economy. With the continuous progress and development of the society, human production activities have an increasing demand for the fossil energy, and the resulting climate change has an increasingly obvious negative impact on the earth's environment. The greenhouse effect is a phenomenon of rising surface temperature caused by the warming effect of greenhouse gases on the earth's surface, which has become one of the main problems threatening the survival of organisms on the earth. Among the gases causing the greenhouse effect, the excessive emission of CO_2 is the main factor exacerbating the greenhouse effect. According to the World Meteorological Organization, the global greenhouse gas concentration continued to increase in 2022, and the concentration of the greenhouse gas in the atmospheric environment reached 415.7 ppm, 149% of the pre-industrial level. Correspondingly, the global average surface temperature in 2022 increased by 1.13 °C compared with the pre-industrial level.

The melting of glaciers and the rise of the sea level caused by the global warming are seriously affecting the human life in the coastal low-altitude areas. According to the study results of the Intergovernmental Panel on Climate Change of the United

© The Author(s), under exclusive license to Springer Nature Switzerland AG 2024
Z. Li et al., *Hydrophobic Ceramic Membranes for CO₂ Capture*,
SpringerBriefs in Energy, https://doi.org/10.1007/978-3-031-77678-6_1

Nations, when the global average temperature rises more than 2 °C, or the concentration of CO_2 in the atmosphere exceeds 450 ppm, the Earth's climate will undergo irreversible catastrophic changes, and the global temperature will further rise, resulting in the extinction of more than 90% of the species on Earth. Therefore, in order to avoid the destruction of the ecological environment and protect the earth home on which human beings depend for survival, effective measures must be taken to reduce CO_2 emissions.

1.2 Research Status

CO_2 capture by membranes is mainly divided into two processes: the membrane separation technology [1] and the membrane absorption technology [2]. The membrane separation method mainly uses the high selectivity of the membrane to achieve the gas separation through the different diffusion rates of different components of the flue gas in the membrane [3]. The membrane absorption rule combines the advantages of the amine solution absorption (high selectivity) and the membrane technology (flexibility) to achieve the efficient and rapid CO_2 capture [4]. In the process of the membrane absorption, the membrane acts as a barrier between the gas phase and the liquid phase. It provides a reaction site for the gas–liquid contact, and the flue gas is selectively absorbed at the liquid–solid interface after passing through the membrane pore. Since the membrane per unit volume can provide a larger contact area, under the condition of a certain total amount of CO_2 capture, the size of the capture equipment can be greatly reduced and the investment cost of the equipment can be reduced [5]. Compared with the conventional amine solution absorption technology such as the packed tower, the membrane technology can independently control the gas and liquid flow to avoid the solvent loss, the channeling, the flue gas entrainment and other problems. In addition, the membrane contactor is light in the weight, low in the operation energy consumption, high CO_2 capture efficiency, low in the environmental impact, easy to scale up the process, and flexible in the installation design. It is very favorable for large-scale industrial applications, and is a CO_2 capture process with good application prospects [6].

In the nature, the superhydrophobicity of animal and plant bodies can be seen everywhere. For example, water droplets roll freely on the surface of a lotus leaf to keep the surface clean all the time [7], water striders walk stably and quickly on the water surface, butterflies can fly freely in a humid environment or even in rainy days [8]. When the contact angle of the droplet is greater than 150° on the surface of the material and the rolling angle is less than 10°, the surface is called a superhydrophobic surface [9, 10]. Inspired by natural superhydrophobic phenomena, researchers have successfully prepared superhydrophobic materials and surfaces for a variety of applications, such as the anti-icing [11], the anti-fouling [12], the self-cleaning [13], the membrane distilled water treatment [14], and the oil–water separation [15]. At present, the common preparation methods of the superhydrophobic mainly include two aspects: First, the construction of special micro and nano structures on the solid surface. Second, the introduction of low surface energy substances on the surface. Zhu et al. [13] sprayed copper nanoparticles on the polypropylene fabric by the spraying method, and obtained the superhydrophobic surface after vacuum drying for 4 h without adding any surface modifier, showing excellent waterproof and self-cleaning properties. Ejeta et al. [15] successfully prepared a superhydrophobic and super-oleophilic material using the cotton as the raw material, and applied it in the oil–water emulsion separation, showing excellent separation performance. Chauhan et al. [16] soaked the cotton fabric in the hexadecyl trimethoxysilane solution by a simple soaking method, and obtained superhydrophobic materials with self-cleaning, stain resistances and antibacterial properties. Zakaria et al. [17] modified calcium carbonate nanoparticles with the stearic acid, and then grafted them on the PVDF membrane surface by 3D imprinting method to construct a porous surface with large pores, and the obtained membrane contact angle could reach up to 151.3°. Zhu et al. [18] obtained an ultra-low surface energy surface through the cross-linking reaction between the fluorosilane and the polydimethylsiloxane, and prepared a hydrophobic membrane. According to the above analysis, it is difficult to make the surface of the superhydrophobic ceramic membrane by grafting only the material with low surface energy. For the preparation of the superhydrophobic ceramic membrane, it is necessary to construct special micro and nano structures on its surface to increase its surface roughness and reduce its surface energy. The sol–gel method and the laser etching technology are used in the construction of micro and nano structures. In the aspect of reducing the surface energy, the impregnation and the chemical vapor deposition are used. Wei et al. [19] constructed a special zinc oxide micro-nano composite structure on the silicon carbide ceramic membrane, and finally grafted the surface with n-octyltriethoxysilane to successfully obtain the superhydrophobic silicon carbide ceramic membrane. Zhang et al. [20] prepared a durable superhydrophobic-superlipophilic composite coating on the surface of the ceramic membrane by the in-situ growth of Al_2O_3 and then the impregnation of polydimethylsiloxane. Hu et al. [21] prepared a high-performance superhydrophobic coating composed of mullite fiber, graphene oxide and methylsiloxane through the 1D/2D hybrid strategy by spraying and heat treatments. The obtained coating remained superhydrophobic and showed good heat resistance after soaking in NaCl, HCl and NaOH aqueous solutions for 200 days. It can maintain superhydrophobicity for 10 h at temperatures up to 350 °C.

Unlike the separation of CO_2 by increasing the pressure difference between the two sides of the membrane in the process of the membrane separation. The membrane in the membrane absorption is not selective, and it needs to be paired with a suitable absorber to selectively absorb CO_2. The common absorbents mainly include the alcohol amine solution [22], the strong alkali solution [23] and the hot potassium alkali solution [24]. In the absorption process, the monoethanolamine (MEA) is the most mature absorber, and researchers have conducted a lot of studies on it, applied in various engineering demonstrations [25].

In addition to membrane materials and absorbers, membrane contactors provide sufficient reaction interfaces for gas and liquid absorption processes. The structure of the membrane contactor directly affects the flow form of gas phase, liquid phase, and mass transfer processes, which has a great influence on the performance of the CO_2 capture. In general, when the membrane contactor adopts the cross-flow flow form, the mass transfer efficiency is higher than that of the membrane contactor with the parallel flow [26]. The shell side fluid passes over the membrane tube in a direction perpendicular to the membrane tube, which is conducive to reducing the fluid cross-flow and the fluid pressure drop. Kiani et al. [27] explored the absorption performance of CO_2 in the contra flow membrane contactor, and studied the influence of factors such as the gas–liquid flowrate, the CO_2 inlet concentration and the membrane component length on the absorption process. The CO_2 capture efficiency increased with the increase of the absorbent concentration, the absorber flowrate and the component length. Moreover, it decreased with the increase of the gas flowrate.

1.3 Description of the Book

Figure 1.1 shows the chapter layout structure of the book. The main content of this book is introduced as follows: In this chapter, the research background and the application fields of the membrane absorption technology are introduced. Chapter 2 mainly introduces the main experimental materials and equipment needed for the research of this study, and also introduces the structure characterization and mass transfer performance testing methods of membrane materials in detail. In Chap. 3, the CO_2 capturing properties of Al_2O_3 ceramic membranes and PTFE membranes are compared. In Chap. 4, the hydrophobic Al_2O_3 ceramic membrane is successfully prepared by grafting FC16 molecules on the commercial micron Al_2O_3 ceramic membrane, and the crystal phase composition, surface morphology, pore size distribution, flux and thermal stability of the ceramic membrane before and after modification are characterized in detail. In Chap. 5, two kinds of Al_2O_3 ceramic membranes with different thickness are successfully modified from initial hydrophilic to superhydrophobic by sol–gel method. Chapter 6 develops porous ceramic membrane based on fly ash, which is a common solid waste in coal-fired power plants, and modifies its surface with fluorosilane solution to obtain a super hydrophobic ceramic membrane based on fly ash. In Chap. 7, the primary concluding remarks of the present book are summarized.

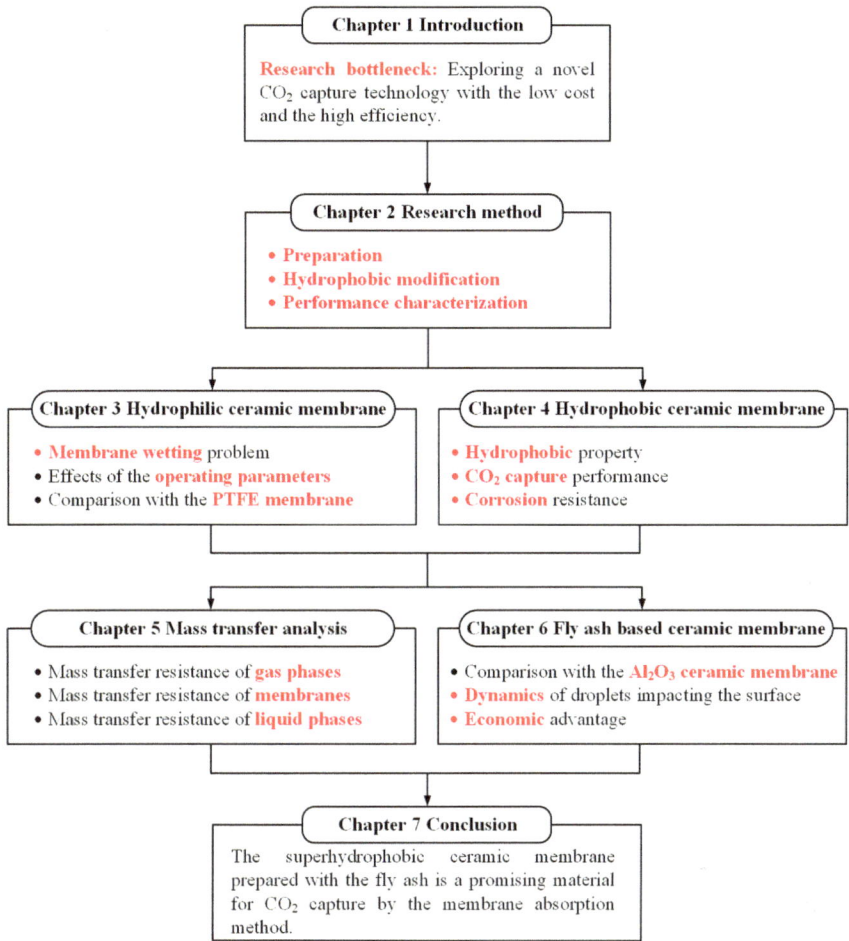

Fig. 1.1 Structure of the present book

References

1. Rahimalimamaghani A, Godini HR, Mboussi M et al (2022) Carbon molecular sieve membranes for selective CO_2 separation at elevated temperatures and pressures. J CO_2 Util 68:102378
2. Zhao Q, Jiang HC, Wang CC et al (2024) Facile fabrication of FAS-PDMS/PEEK composite hollow fiber membrane with honeycomb-like structure for CO_2 capture from flue gas by membrane absorption. Sep Purif Technol 329:124845
3. Han Y, Ho WSW (2021) Polymeric membranes for CO_2 separation and capture. J Membr Sci 628:119244
4. Rosli A, Ahmad AL, Low SC (2020) Enhancing membrane hydrophobicity using silica end-capped with organosilicon for CO_2 absorption in membrane contactor. Sep Purif Technol 251:117429

5. Lu SS, Huang C, Shen YD et al (2021) Research progress of membrane contactor on CO_2 capture. Environ Chem 40:1088–1099 (in Chinese)
6. Kim S, Heath DE, Kentish SE (2022) Improved carbon dioxide stripping by membrane contactors using hydrophobic electrospun poly(vinylidene fluoride-co-hexafluoro propylene) (PVDF-HFP) membranes. Chem Eng J 428:131247
7. Lin F, Li SH, Li YS (2002) Super-hydrophobic surfaces: from natural to artificial. Adv Mater 14:1857–1860
8. Zheng YM, Gao XF, Jiang L (2007) Directional adhesion of superhydrophobic butterfly wings. Soft Matter 3:178–182
9. Wang ZX, Elimelech M, Lin SH (2016) Environmental applications of interfacial materials with special wettability. Environ Sci Technol 50:2132–2150
10. Zhang WL, Wang DH, Sun ZN et al (2016) Robust superhydrophobicity: mechanisms and strategies. Chem Soc Rev 50:4031–4061
11. Wang N, Xiong DS, Deng YL et al (2015) Mechanically robust superhydrophobic steel surface with anti-icing, UV-durability, and corrosion resistance properties. ACS Appl Mater Interfaces 7:6260–6272
12. Li H, Yu SR, Hu JH et al (2019) Modifier-free fabrication of durable superhydrophobic electrodeposited Cu-Zn coating on steel substrate with self-cleaning, anti-corrosion and anti-scaling properties. Appl Surf Sci 481:872–882
13. Zhu SM, Kang ZX, Wang F et al (2021) Copper nanoparticle decorated non-woven polypropylene fabrics with durable superhydrophobicity and conductivity. Nanotechnology 32:035701
14. Kujawa J, Kujawski W, Cerneaux S et al (2020) Zirconium dioxide membranes decorated by silanes based-modifiers for membrane distillation-Material chemistry approach. J Membr Sci 596:117597
15. Ejeta DD, Wang CF, Kuo SW et al (2020) Preparation of superhydrophobic and superoleophilic cotton-based material for extremely high flux water-in-oil emulsion separation. Chem Eng J 402:126289
16. Chauhan P, Kumar A, Bhushan B (2019) Self-cleaning, stain-resistant and anti-bacterial superhydrophobic cotton fabric prepared by simple immersion technique. J Colloid Interf Sci 535:66–74
17. Zaliman SQ, Zakaria NA, Ahmad AL et al (2022) 3D-imprinted superhydrophobic polyvinylidene fluoride membrane contactor incorporated with $CaCO_3$ nanoparticles for carbon capture. Sep Purif Technol 287:120519
18. Zhu HP, Li XR, Pan Y et al (2020) Fluorinated PDMS membrane with anti-biofouling property for in-situ biobutanol recovery from fermentation-pervaporation coupled process. J Membr Sci 609:118225
19. Wei YB, Xie ZX, Qi H (2020) Superhydrophobic-superoleophilic SiC membranes with micro-nano hierarchical structures for high-efficient water-in-oil emulsion separation. J Membr Sci 601:117842
20. Zhang MD, Ning HL, Shang JJ et al (2024) A robust superhydrophobic-superoleophilic PDMS/Al2O3/CM composite ceramic membrane: stability, efficient emulsified oil/water separation, and anti-pollution performance. Sep Purif Technol 328:124864
21. Hu XB, Lu CB, Yang BS et al (2024) Ecofriendly fabrication of superhydrophobic coating with high chemical/mechanical durability and enhanced heat resistance by 1D/2D hybridization strategy. Colloid Surface A 685:133279
22. Oh HT, Lee JC, Lee CH (2022) Performance and sensitivity analysis of packed-column absorption process using multi-amine solvents for post-combustion CO_2 capture. Fuel 314:122768
23. Rastegar Z, Ghaemi A (2022) CO_2 absorption into potassium hydroxide aqueous solution: experimental and modeling. Heat Mass Transf 58:365–381
24. Zhang LQ, Gai XK, Yang RQ et al (2017) Kinetic study of absorption of carbon dioxide in biogas using hot potassium carbonate solution process in a dual-driver reactor. Acta Sci Circum 37:2490–2497 (in Chinese)

25. Sun PX, Jia Y, Qi CJ et al (2024) Synergistic promoting of CO_2 absorption-mineralization by MEA-carbide slag. Sep Purif Technol 341:126899
26. Mansourizadeh A, Ismail AF (2009) Hollow fiber gas-liquid membrane contactors for acid gas capture: a review. J Hazard Mater 171:38–53
27. Kiani S, Taghizade A, Ramezani R et al (2022) Enhancement of CO_2 removal by promoted MDEA solution in a hollow fiber membrane contactor: a numerical and experimental study. Carbon Capture Sci Technol 2:100028

Chapter 2
Research Methods

This chapter mainly introduces the main experimental materials and equipment needed for the research of this study, and also introduces the structure characterization and mass transfer performance testing methods of membrane materials in detail, which lays a foundation for the research of subsequent chapters.

2.1 Preparation of Ceramic Composite Membranes

The experimental materials used in this book are shown in Table 2.1. Furthermore, the main experimental instruments and equipment involved in this book are shown in Table 2.2.

2.2 Characterization of Ceramic Composite Membranes

Scanning electron microscope uses a narrow focus high-energy electron beam to scan the sample, and generates various physical information through the interaction between the electron beam and the substance. By collecting, magnifying and re-imaging these information, the microstructure of the substance can be characterized. In this book, the microstructure and the surface morphology of the surface and the cross section are observed by the scanning electron microscope (S-4800, Hitachi, Japan). When observing the microstructure of the membrane section, the ceramic membrane is directly broken by the external force, and the PTFE membrane is quenched by the liquid nitrogen to avoid cutting the broken membrane section morphology. Due to the poor conductivity of the membrane, in order to improve the conductivity, the surface and cross-section of all samples are treated with the gold spraying for 60 s before the experiment. In addition, the elemental composition

Table 2.1 Materials used in the experiment

Materials	Specification	Manufacturer
Ceramic membrane with a pore size of 10 nm	/	Qiangrui, Hefei, China
Ceramic membrane with a pore size of 0.2 μm (M1)	/	Jingang, Foshan, China
Ceramic membrane with a pore size of 0.2 μm (M2)	/	Sinyuan, Ningbo, China
PTFE hollow fiber membrane	/	Keborui, Hefei, China
Fly ash	/	Tokto Thermal Power Plant, Huhhot, China
Ethanolamine	Analytical reagent	Aladin, Shanghai, China
Anhydrous ethanol	99.9%	Xindaidongfang, Beijing, China
Concentrated sulfuric acid	98%	Beijing Chemical Works, Beijing, China
Sodium hydroxide	Analytical reagent	Beijing Chemical Works, Beijing, China
Anhydrous methanol	Analytical reagent	Beijing Chemical Works, Beijing, China
Acetone	Analytical reagent	Aladin, Shanghai, China
Ethyl orthosilicate	Analytical reagent	Aladin, Shanghai, China
Ammonia liquor	35 vol%	Aladin, Shanghai, China
Carboxymethyl cellulose	Analytical reagent	Lihong, Chongqing, China
Dextrin	Analytical reagent	Lihong, Chongqing, China
Glycerin	EP	Aladin, Shanghai, China
1H, 1H, 2H, 2H-perfluorodecyl triethoxysilane	96%	Aladin, Shanghai, China
Deionized water	5 μS/cm	Self-made

of the membrane surface is analyzed by the scanning electron microscope-energy dispersion spectrometry.

An atomic force microscope is an instrument used to study the surface structure of solid materials. By detecting the extremely weak interatomic interaction force between the sample surface and the micro force sensor, the surface structure and properties of the substance are reflected. When scanning the sample, the sensor detects the change of the interaction force, and the distribution information of the interaction force can be obtained. Therefore, the surface morphology, the structure and the surface roughness information can be obtained with the nanometer resolution. The surface morphology and the roughness of original and modified ceramic membranes are observed and measured by the atomic force microscopy (Dimension Icon, Bruker, Germany) in the contact mode.

The contact angle and the rolling angle of the membrane are measured by a contact angle measuring instrument (Kruss-Advsncse, Germany). The test method is set up

Table 2.2 Equipment used for the membrane characterization

Equipment	Model	Manufacturer country	Application
Scanning electron microscope	S4800	Japan	Observing the surface and cross section morphology of the membrane
Contact angle measuring instrument	KRUSS	Germany	Measuring the contact angle of the membrane surface
Aperture analyzer	PSDA-30	China	Measuring the N_2 flux and the pore size distribution
Water flux test equipment	Self-made	China	Measuring the pure water flux of the membrane
Atomic force microscope	Dimension Icon	Germany	Measuring the surface roughness of the membrane
Fourier transform infrared spectrum analyzer	Thermo Nicolet iS5	United States	Measuring the chemical composition of the membrane surface
X-ray diffractometer	Al-Y3000	China	Measuring the crystal phase composition of the ceramic membrane
Thermogravimetric analyzer	Sta 449 F5	Germany	Measuring the thermal stability of the ceramic membrane
X-ray fluorescence analyzer	Zetium	Netherlands	Testing the elemental composition of the fly ash
High-speed camera	X113	China	Observing the dynamic characteristics of droplets impacting on the ceramic membrane surface
CO_2 concentration analyzer	GT1000-CO_2-WL	China	Measuring the concentration of CO_2

with the test solvent being the deionized water, the ethanolamine solution or the other liquid solution, and the drop volume is 3 μL. The measurement process is carried out at the room temperature, and the surface contact angle is measured five times at different positions on the membrane surface, and the measurement error is $\pm 1°$. When measuring the rolling angle of the flat membrane, the droplet is first dropped on the surface of the flat membrane, and then the test bench is slowly tilted until the droplet just falls from the surface to stop, and the angle between the test bench and the horizontal plane is the rolling angle of the membrane.

The liquid breakthrough pressure (LEP) is an important parameter to characterize the hydrophobicity of the membrane. When the pressure difference between the two sides of the membrane is lower than the LEP, the membrane cannot easy to be wetted. When the pressure difference between the two sides of the membrane is higher than the LEP, the liquid can easily pass through holes and wet the membrane. The larger the LEP, the more difficult the membrane is to wet. Theoretical LEP is calculated according to Eq. (2.1).

$$LEP = -\frac{2\gamma_L}{r_{max}}\cos\theta \tag{2.1}$$

where "γ_L" represents the surface tension of the liquid. "r_{max}" represents the maximum radius of ceramic membrane pores. "θ" represents the contact angle of the liquid at the membrane surface.

The chemical composition of the membrane surface is measured by the Fourier transform infrared spectroscopy, which is used to analyze the chemical composition of the ceramic membrane surface before and after the hydrophobic property. Furthermore, it can be utilized to judge the success or failure of the graft modification of the ceramic membrane. In this book, the surface composition of ceramic membrane M1 and M2 is analyzed by the Fourier transform infrared spectroscopy (Thermo Nicolet iS5, USA), and 16 scans are performed in the range of 4.0 cm, and the scanning wavelength range is 400–4000 cm^{-1}.

The crystal phase composition of the fly ash and the ceramic membrane is obtained by the X-ray Diffraction (Al-Y3000, Aolong, Dandong, China). The diffraction process uses Cu-Kα radiation ($\lambda = 0.154$ nm), scanning voltage is 30 kV, scanning current is 25 mA, scanning angle is $10°$–$90°$, and the scanning rate is $4°$/min. MDI Jade 6.0 software is used to analyze the diffraction data and analyze the crystal phase composition of the sample.

In order to verify the thermal stability of the ceramic membrane, a thermogravimetric analysis is performed on the ceramic membrane (Sta 449 F5, Netzsch, Germany). In the N_2 environment, the mass change of the sample to be measured by continuously increasing the temperature, the temperature rise rate is 10 °C/min, and the test temperature range is from the room temperature to 800 °C.

The pore size distribution and the N_2 flux of ceramic membranes are measured by the aperture analyzer (PSDA-30, Gaoqian, Nanjing, China). In the pore size test, the ceramic membrane is first placed in a vacuum infiltrator and left for 5 min in a vacuum environment of -100 kPa. As a result, the ceramic membrane is completely infiltrated with the infiltrating fluid of QG16 with a surface tension of 16 mN/m. Then, the high purity N_2 is used as the gas source, and the pore size distribution is measured in an aperture analyzer. At the same time, the N_2 permeation flux of the ceramic membrane is measured by the dry curve method.

The porosity of the ceramic membrane is measured by the Archimedes method, namely gravimetric method. The details are as follows. First, the ceramic membrane sample is placed in a vacuum drying oven and dried at 100 °C for 3 h. After cooling to the room temperature, the mass of the ceramic membrane sample is measured and recorded as m_1. Second, the ceramic membrane sample is boiled in the deionized water for 3 h to ensure that the ceramic membrane is completely infiltrated and cooled to the room temperature. Third, adding the deionized water to the beaker and placing it on the balance, and setting the indicator to zero. Forth, a fine copper wire is used to suspend the ceramic membrane on the beaker, ensuring that it does not touch the wall of the beaker. At this time, the balance reading is the floating weight of the ceramic membrane sample, recording as m_2. Finally, the submerged ceramic membrane sample is taken out, the water on the outer surface of the sample is wiped dry, and the mass of the sample at this time is quickly weighed. The weight is recorded as m_3, that is, the wet weight of the sample. As a result, the porosity is calculated by Eq. (2.2).

$$\varepsilon = \frac{m_3 - m_1}{m_3 - m_2} \times 100\% \tag{2.2}$$

where "ε" represents the porosity.

The pure water flux of the ceramic membrane is determined by the self-made equipment, which is shown in Fig. 2.1. The ceramic membrane to be tested is first placed in a self-made assembly with one end sealed and the other end connected to the water tank. During the experiment, the pressure reducing valve of the nitrogen cylinder is first opened, and the pressure difference between inside and outside the membrane is controlled by the opening of the valve. At the same time, the mass of the permeated water is measured by an electronic balance, and the volume of the permeated water is obtained after the conversion. The water flux of the ceramic membrane can be calculated by Eq. (2.3).

$$J_0 = \frac{VA}{\Delta t} \tag{2.3}$$

where "J_0" represents the pure water flux. "V" represents the volume of the permeated water. "A" represents the membrane area. "Δt" represents the experimental time.

X-ray fluorescence analysis uses the interaction of X-rays with a material to determine its elemental composition. In this book, X-ray fluorescence diffraction (Zetium, Netherlands) is used to measure and calculate the elemental composition of the fly ash used.

In order to explore the interaction between droplets and the surface of ceramic membrane, an experimental system for droplets impacting on the membrane surface is built, which is shown in Fig. 2.2. The experimental system is mainly composed of the high-speed camera (China Fuhuang Junda, X113), the iron frame, the micro-syringe, the test tube clip, and the LED light source. The specific experimental process is as follows. Using a micro syringe to push slowly at a certain rate, a liquid drop with a diameter of about 2 mm is generated at the tip of the needle, and it is made to do free fall. The falling velocity of the liquid drop can be changed by adjusting the height of the tube clip. A high-speed camera is used to record the instantaneous dynamic process of droplets impacting on the membrane surface, and the shooting

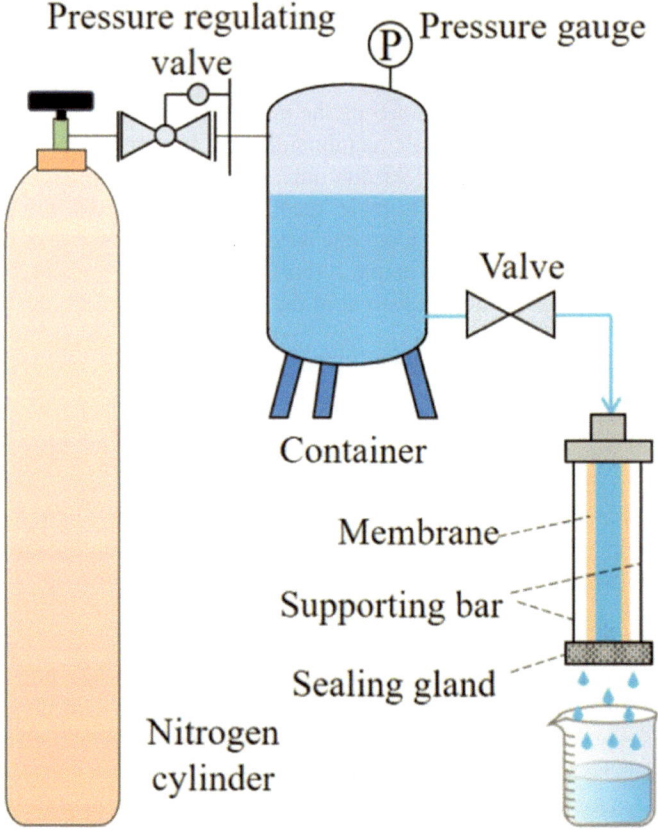

Fig. 2.1 Schematic diagram of the equipment for the water flux measurement. Reprinted with the permission from Ref. [1] Copyright (2022) (Elsevier)

speed is 6000 f/s. The captured images are saved and processed by the computer, and each experiment process is repeated three times.

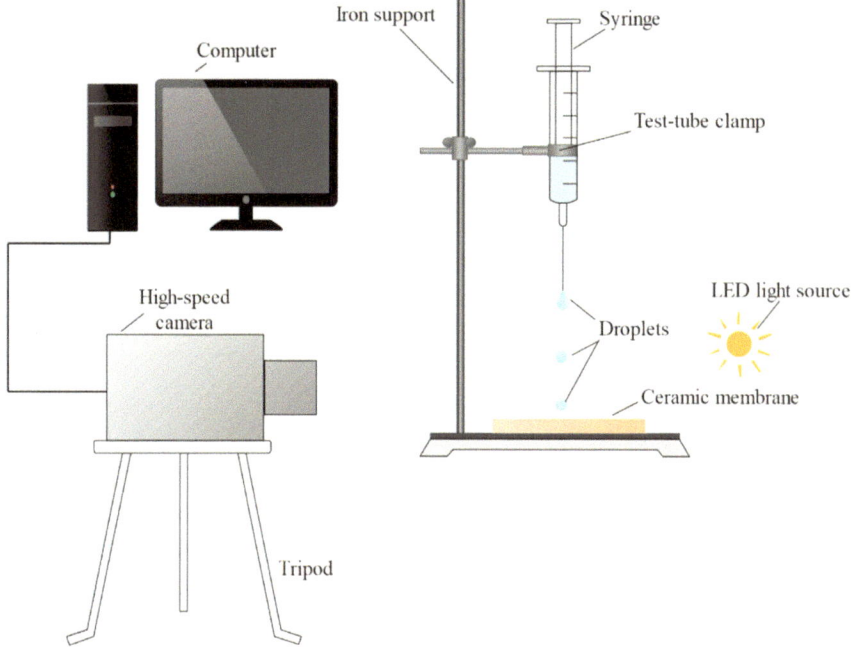

Fig. 2.2 Schematic diagram of the high-speed camera experimental device. Reprinted with the permission from Ref. [2] Copyright (2022) (Elsevier)

References

1. Huang JG, Zhang H, Zhang YT et al (2022) Recycle coal fly ash for preparing tubular ceramic membranes applied in transport membrane condenser. Sep Purif Technol 282:119972
2. Fu HM, Li ZH, Zhang YT et al (2022) Preparation, characterization and properties study of a superhydrophobic ceramic membrane based on fly ash. Ceram Int 48:11573–11587

Chapter 3
Analysis of CO_2 Mass Transfer Performance of Membrane Contactors

In this chapter, the effects of different operating parameters on the CO_2 capture performance using the ceramic membrane with a pore size of 10 nm are studied experimentally. In addition, a detailed comparison is made with PTFE polymer membrane in terms of the membrane material, the system operation mode and the CO_2 capture performance. It can further prove the feasibility of hydrophilic ceramic membranes in the CO_2 capture, and provide an alternative technical idea for the membrane technology to capture CO_2 in practical engineering applications.

3.1 Experimental Platform

Figures 3.1 and 3.2 respectively show the schematic diagram and the actual diagram of the CO_2 capture experimental platform of the membrane contactor. The experimental platform is mainly composed of the gas side, the liquid side and the membrane side. The gas side is mainly composed of a high-pressure mixed gas cylinder (N_2:CO_2 = 4:1), a pressure reducing valve, a gas pressure gauge, a gas flow meter, a beaker with the enough NaOH solution, an electronic balance, and pipes and valves. The pressure of simulated flue gas can be adjusted by controlling the pressure reducing valve and the membrane contactor outlet valve. By controlling the valve of the gas flowmeter, the flowrate can be adjusted. In the experiment, the gas flowrate is first adjusted, and then the gas pressure is adjusted. The simulated flue gas flows out through the gas side and membrane contactor, and is finally treated with the concentrated NaOH solution and then discharged into the air. At the end of the experiment, the added mass of the NaOH solution is the mass of the uncaptured CO_2 in the simulated flue gas. The liquid side of the experimental system is composed of a lean liquid tank with heater, a booster pump (maximum flowrate is 10 L/min, and maximum head is 14 m), pressure gauges, liquid flowmeters, digital thermometers, a rich liquid tank, pipelines and valves. The solution flowrate can be adjusted by controlling the inlet valve of the

© The Author(s), under exclusive license to Springer Nature Switzerland AG 2024   17
Z. Li et al., *Hydrophobic Ceramic Membranes for CO2 Capture*,
SpringerBriefs in Energy, https://doi.org/10.1007/978-3-031-77678-6_3

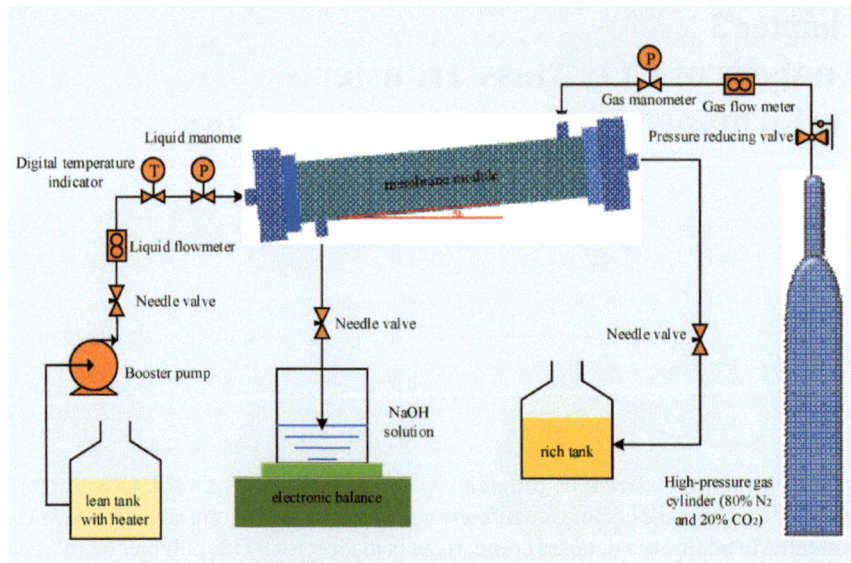

Fig. 3.1 Schematic diagram of the experimental platform for investigating the membrane contactor. Reprinted with the permission from Ref. [1] Copyright (2023) (Elsevier)

membrane contactor. The solution pressure is regulated by controlling the valve at the outlet of the membrane contactor. In the experiment, first adjusting the solution flowrate, and then adjusting the solution pressure. In addition, in order to ensure the stable operation of the system, the liquid side circulation should be opened during the experiment, and the gas side circulation should be opened after the liquid side circulation is stable. The measuring equipment involved in this experiment platform is shown in Table 3.1.

During the experiment, the CO$_2$ capture performance of a ceramic membrane with a pore size of 10 nm and a PTFE membrane contactor with a pore size of 0.1 μm is compared. The structure of the PTFE membrane contactor is shown in Figs. 3.3a and 3.4a. It is composed of 1050 PTFE membranes with an outer diameter of 1 mm and an inner diameter of 0.52 mm. The internal and external structure of the ceramic membrane contactor is shown in Figs. 3.3b and 3.4b, which consists of a ceramic membrane with an outer diameter of 12 mm and an inner diameter of 8 mm. Membrane areas of the ceramic membrane contactor and the PTFE membrane contactor are 0.02 m^2 and 0.8 m^2, respectively. Table 3.2 lists the detailed parameters of both membrane contactors. During the experiment, the membrane contactor is placed at an angle of 5° from the horizontal. Therefore, the absorbent can fill the entire membrane contactor. The control variable method is used in the experiment. Since the gas–liquid membrane contactor can obtain the high gas mass transfer efficiency when it flows in the counter-current mode, all experiments in this chapter adopt the gas–liquid counter-current mode. Table 3.3 lists the parameters used in the experiment.

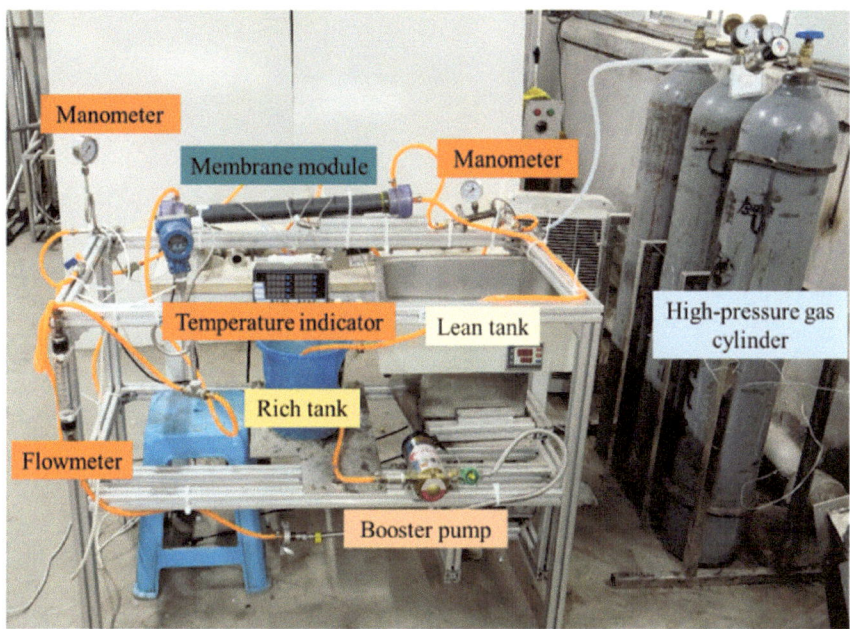

Fig. 3.2 Photograph of the experimental platform for investigating the membrane contactor. Reprinted with the permission from Ref. [1] Copyright (2023) (Elsevier)

Table 3.1 Equipment in experiments of the membrane contactor

Equipment	Manufacturer	Model	Precision (%)
Digital gas flowmeter	Sevenstar Electronics, Beijing, China	DO7-9E	±2.0
Liquid flowmeter	Chunhui, Chuzhou, China	M-15	±4.0
Digital thermometer	Xinrui, Changzhou, China	HH-6PT	±0.17
Gas pressure gauge	Jiangyue, Shanghai, China	Y-60BF	±2.5
Liquid pressure gauge	Hongqi, Wenzhou, China	Y-100	±1.6
Electronic balance	Jiming, Yuyao, China	JM-A20002	±0.01

3.2 Evaluation Index of Performance

The CO_2 capture efficiency and the CO_2 mass transfer rate are used to evaluate the CO_2 capture performance of membrane contactors. The calculation methods are shown in Eqs. (3.1) and (3.2) [2].

$$\eta = 1 - \frac{\Delta m}{Q_{g,in}\varphi\rho\,\Delta t} \qquad (3.1)$$

(a) PTFE membranecontactor.

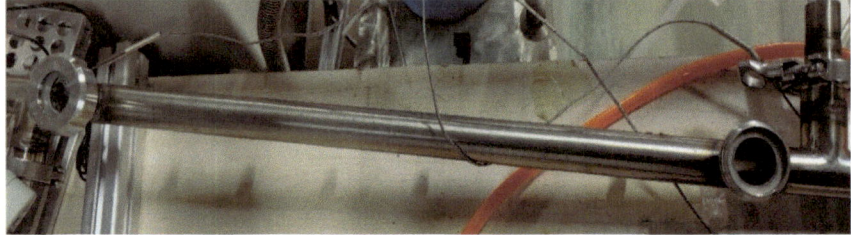

(b) Ceramic membranecontactor.

Fig. 3.3 Photograph of membrane contactors. Reprinted with the permission from Ref. [1] Copyright (2023) (Elsevier)

(a)   (b)

Fig. 3.4 Internal structure of membrane contactors (**a** PTFE membrane contactor; **b** ceramic membrane contactor). Reprinted with the permission from Ref. [1] Copyright (2023) (Elsevier)

$$J = \frac{Q_{g,in}\varphi\rho - \Delta m}{2640A} \tag{3.2}$$

where "η" represents the CO$_2$ capture efficiency. "Δm" represents the mass difference of the NaOH solution before and after experiments. "$Q_{g,in}$" represents the inlet gas flowrate in the membrane contactor. "φ" represents the CO$_2$ volume fraction in the flue gas. "ρ" represents the CO$_2$ density. "J" represents the mass transfer rate.

Table 3.2 Parameters of PTFE and ceramic membrane contactors

Items	PTFE membrane contactor	Ceramic membrane contactor
Material	PTFE	Al_2O_3
Outer diameter of the membrane (mm)	1	12
Inner diameter of the membrane (mm)	0.52	8
Length of the membrane (mm)	500	800
Number of the membrane	1050	1
Outer diameter of the membrane contactor (mm)	50	40
Inner diameter of the membrane contactor (mm)	42	30
Membrane area (m^2)	0.8	0.02
Average pore size (nm)	100	10
Porosity (%)	55	43

Table 3.3 Parameter ranges of CO_2 capture experiments

Parameters	Ranges
Gas flowrate	2.5–18.1 L/min
Absorbent flowrate	0.2–0.65 L/min
Gas pressure	0.01–0.04 MPa
Absorbent temperature	293.15–338.15 K

3.3 Material Property Analysis

Figure 3.5 shows scanning electron microscope morphologies of the ceramic membrane and the PTFE membrane. The pore distribution on the surface of the ceramic membrane is uniform, and there are no obvious large pore defects and cracks. It can be seen from the cross-sectional morphology that the ceramic membrane has a typical asymmetric structure, consisting of a supporting layer containing large pores and a selective layer containing nanopores. The supporting layer accounts for the majority of the whole ceramic membrane and provides a certain strength for the whole supporting performance of the ceramic membrane. The thickness of the nano-layer is only about 10 µm, which provides a certain selectivity for the ceramic membrane. Compared with the ceramic membrane, the PTFE membrane surface is smooth, its cross section structure is the filament-like structure. Moreover, the structure can provide the sufficient specific surface area for the reaction of CO_2 and absorbents, accelerate the gas mass transfer process, and improve the mass transfer efficiency of the membrane contactor.

Figure 3.6 shows measurement results of contact angles. The contact angle of the ceramic membrane is significantly smaller than that of the PTFE membrane,

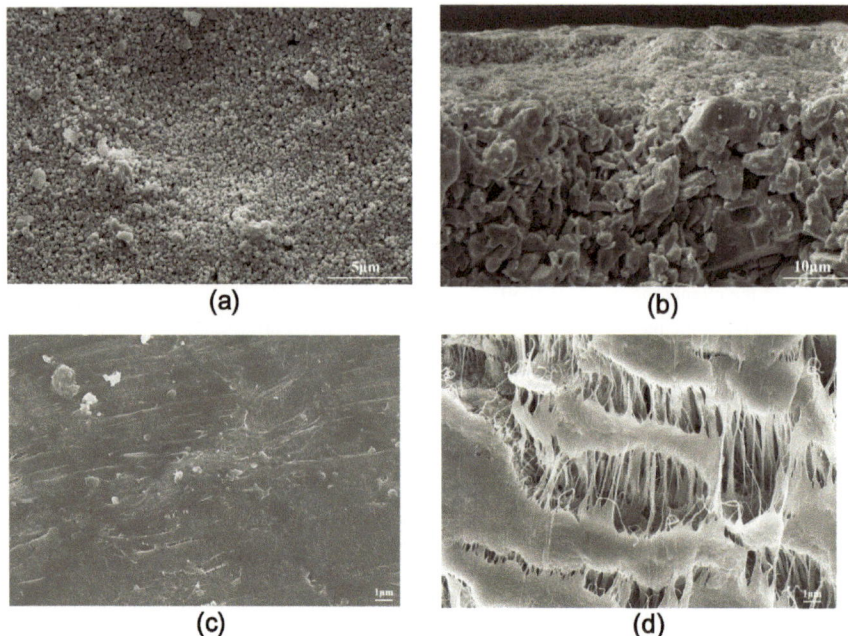

Fig. 3.5 Surface and cross-sectional scanning electron microscope images for ceramic and PTFE membranes (**a** ceramic membrane surface; **b** ceramic membrane section; **c** PTFE membrane surface; **d** PTFE membrane section). Reprinted with the permission from Ref. [1] Copyright (2023) (Elsevier)

which is due to the lower surface energy of the PTFE membrane and its higher hydrophobicity. The contact angle of the ceramic membrane is 87.5°, showing a certain hydrophilicity. While the contact angle of the PTFE membrane is 112.3°, showing the good hydrophobicity.

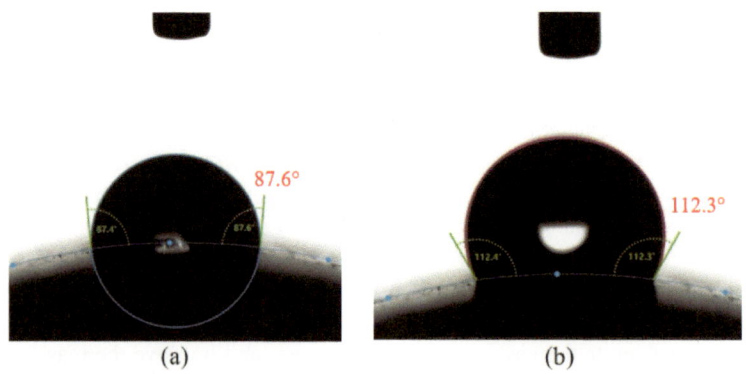

Fig. 3.6 Contact angles of ceramic and PTFE membranes (**a** ceramic membrane; **b** PTFE membrane)

Figure 3.7 shows the variation of contact angles. Due to the porous structure of the ceramic membrane, droplets quickly penetrate into the ceramic membrane after touching the surface, and the contact angle decreases continuously during this process. In contrast, due to the hydrophobicity of the PTFE membrane, droplets always stay on the surface after touching the membrane surface and do not penetrate into the membrane. Therefore, the PTFE membrane has better wettability than the ceramic membrane. In the actual CO_2 capture process, adjusting the operation mode of the system according to the different characteristics of the membrane material plays an important role in inhibiting the membrane wetting and maintaining the system stability. In the process of CO_2 capture by the PTFE membrane contactor, due to the good hydrophobicity of the PTFE membrane, the system can be operated with the gas pressure on the membrane side slightly lower than the absorbent pressure on the liquid side. A high CO_2 capture flux can be achieved by ensuring that the pressure difference between both sides of the membrane cannot exceed the PTFE membrane breakthrough pressure. In the ceramic membrane contactor, due to the hydrophilicity of the membrane, the gas pressure on the membrane side is slightly higher than that on the liquid side. Thus, the membrane wetting is avoided and the mass transfer performance of the ceramic membrane contactor is guaranteed.

Figures 3.8 and 3.9 show the changes of fluid fluxes with the pressure in ceramic and PTFE membranes, respectively. Although the pore size of the ceramic membrane is smaller than that of the PTFE membrane, it has a higher gas flux. N_2 fluxes of both membranes increase with the increase of pressures. The N_2 flux of the PTFE membrane increases linearly with the pressure, while that of the ceramic membrane

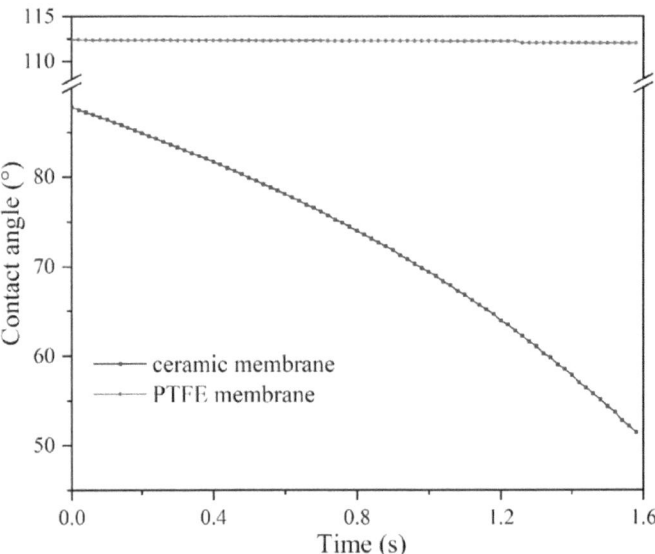

Fig. 3.7 Variation of the contact angle for ceramic and PTFE membranes with the time. Reprinted with the permission from Ref. [1] Copyright (2023) (Elsevier)

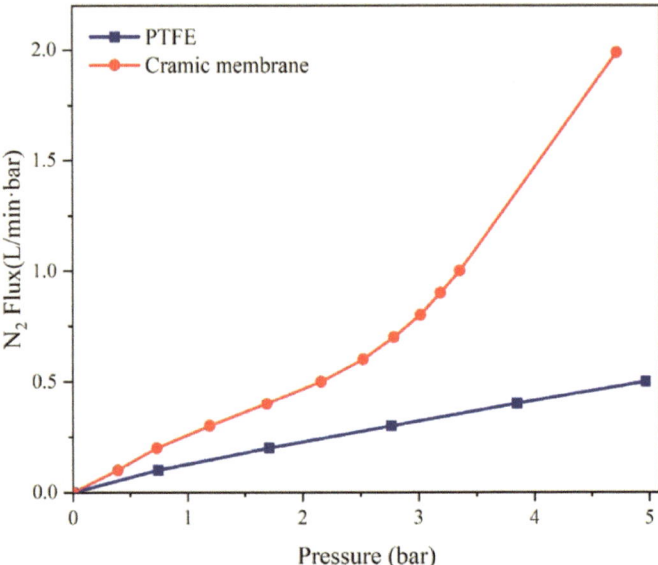

Fig. 3.8 N$_2$ fluxes of ceramic and PTFE membranes change with the pressure. Reprinted with the permission from Ref. [1] Copyright (2023) (Elsevier)

increases slowly at first and then rapidly. The thickness of the ceramic membrane is 2 mm, and the pore size distribution is more complicated in the radial direction. However, the thickness of the PTFE membrane is only 0.24 mm, and the pore size distribution is more uniform in the radial direction. The ceramic membrane has an obvious layered pore size distribution in the radial direction. Therefore, the N$_2$ flux of the ceramic membrane increases nonlinearly with the increase of the pressure. The water flux of the ceramic membrane also increases with the increase of the pressure. For PTFE membranes, the water flux is always 0, even if the pressure increases to 2.5 bar. It means that the PTFE membrane has the excellent wettability and its liquid breakthrough pressure is above 2.5 bar.

3.4 Performance of CO$_2$ Capture

3.4.1 Effect of the Absorbent Flowrate

Figure 3.10 shows effects of the absorbent flowrate. CO$_2$ capture efficiencies and mass transfer rates of membrane contactors increase significantly with the increase of absorbent flowrates. For the ceramic membrane contactor, when the absorbent flowrate increases from 0.2 to 0.6 L/min, the CO$_2$ mass transfer rate increases from 41.83×10^{-3} to 47.95×10^{-3} mol/(m^2 s), and the capture efficiency increases from

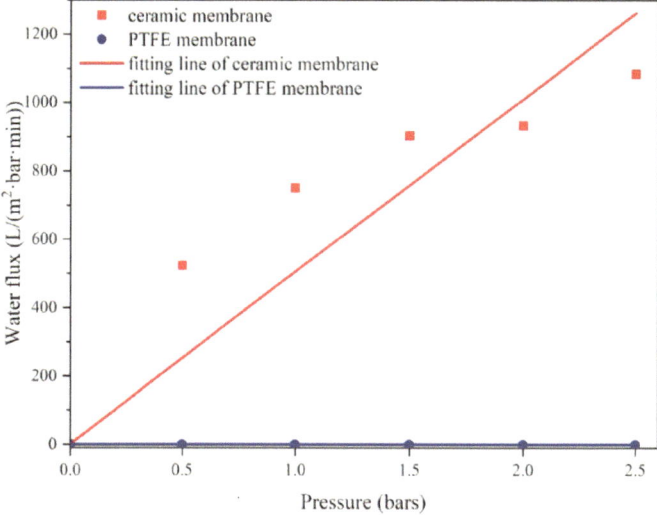

Fig. 3.9 Water fluxes of ceramic and PTFE membranes change with the pressure. Reprinted with the permission from Ref. [1] Copyright (2023) (Elsevier)

69.71 to 77.73%. While for the PTFE membrane contactor, the CO$_2$ mass transfer rate increases from 1.94×10^{-3} to 2.14×10^{-3} mol/(m^2 s), and the capture efficiency increases from 87.06 to 96.89%. With the increase of the absorbent flowrate, the thickness of the liquid–solid boundary layer decreases and the mass transfer resistance decreases. In addition, the turbulence at the membrane interface increases with the increase of the absorbent flowrate. The reaction products between the gas and the absorbent are quickly carried, and the reactants at the membrane interface are quickly updated, which is conducive to the positive reaction between CO$_2$ and MEA absorbent. Comparing the CO$_2$ capture performance of two membrane contactors, the CO$_2$ capture efficiency of the PTFE membrane contactor is higher, while the CO$_2$ mass transfer rate of the ceramic membrane contactor is higher. Generally speaking, large pore size is conducive to the membrane mass transfer process. During the experiment, the average pore size of the selected layer is 10 nm, and the pore size of the support layer is in the range of 1–10 μm. The average pore size of the PTFE membrane is 0.1 μm. For ceramic membranes, the nano-layer thickness is about 10 μm, which is almost negligible compared to the total thickness of the membrane (2 mm) [3]. Therefore, the ceramic membrane has a higher mass transfer rate.

3.4.2 Effect of the Absorbent Temperature

Figure 3.11 shows the effect of the absorbent temperature. With the increase of the temperature, the CO$_2$ capture efficiency and the CO$_2$ mass transfer rate of the PTFE

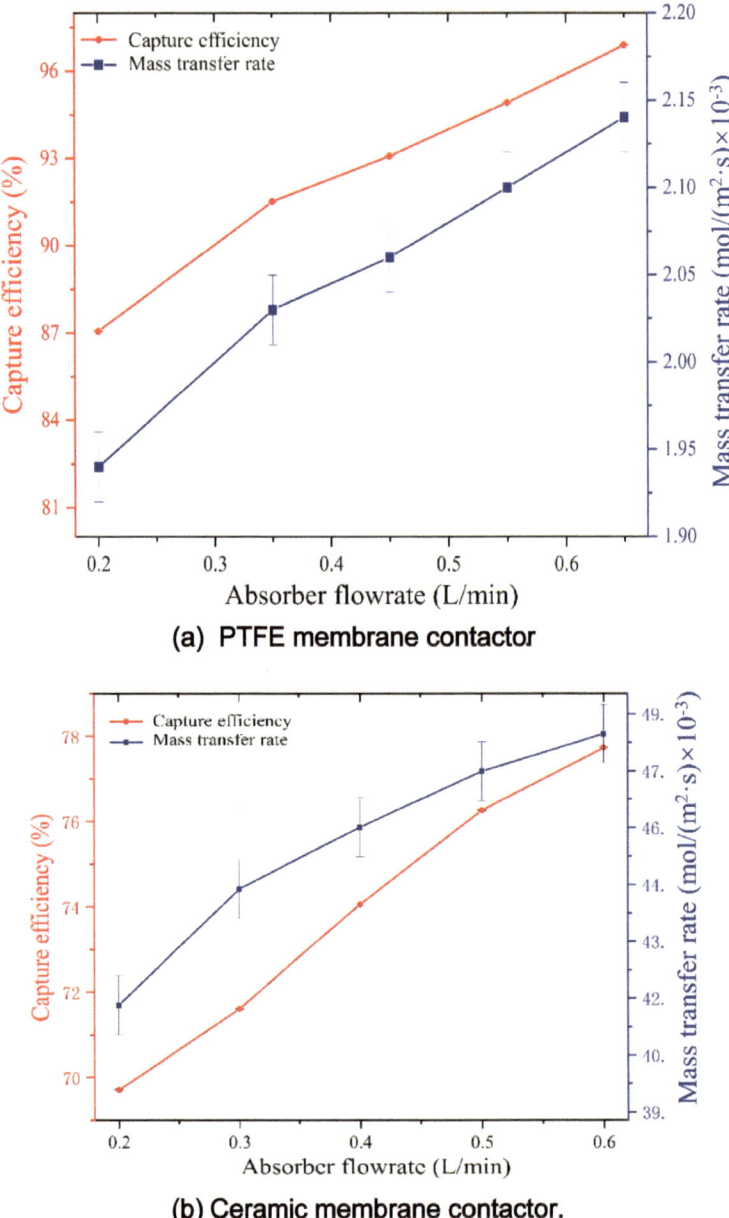

(a) PTFE membrane contactor

(b) Ceramic membrane contactor.

Fig. 3.10 Effect of the absorber flowrate on the capture performance

membrane contactor remain around 90% and 1.11×10^{-3} mol/(m^2 s), respectively. While the CO_2 capture performance and the CO_2 mass transfer rate of the ceramic membrane contactor show an obvious trend of increasing at first and then slowly stabilizing. Increasing the temperature accelerates the reaction velocity, which is conducive to the CO_2 absorption reaction of the MEA solution. However, the positive reaction between CO_2 and MEA is an exothermic reaction, and the rise of temperature promotes the reverse reaction, which is not conducive to the absorption of CO_2. For PTFE membrane contactors, the CO_2 capture efficiency is always high, and the temperature increase has little effect on it. For ceramic membrane contactors, the initial stage of the temperature change is dominant in promoting the reaction, resulting in a significant increase in the CO_2 capture efficiency. When the temperature rises to about 312 K, the reverse reaction rate of MEA and CO_2 reversible reaction increases, limiting the further improvement of the CO_2 capture efficiency, showing a slow and stable trend.

3.4.3 Effect of the Gas Flowrate

Figure 3.12 shows the effect of the gas flowrate. When the gas flowrate increases, the CO_2 mass transfer rate of both PTFE membrane and ceramic membrane contactors increase significantly. Increasing the gas flowrate accelerates the regeneration rate of CO_2 in the flue gas and the gas–liquid interface, thus increasing the reaction rate. Moreover, the thickness of the gas–liquid boundary layer decreases, and the gas mass transfer resistance between the gas phase and the liquid phase decreases [4]. In addition, with the increase of the gas flowrate, the CO_2 capture efficiency of the PTFE membrane contactor decreases, while the CO_2 capture efficiency of the ceramic membrane contactor increases. With a PTFE membrane area of 0.8 m^2 and a ceramic membrane area of 0.02 m^2, the PTFE membrane contactor can provide a higher gas–liquid reaction area. For the PTFE membrane contactor, when the gas flowrate is 5 L/min, the CO_2 capture efficiency is close to 99%. With the increase of the gas flowrate, the CO_2 capture efficiency decreases. It indicates that when the gas flow rate is 5 L/min, the absorption capacity of the solution to CO_2 has reached the limit. Further increasing in the gas flowrate cannot absorb more CO_2, leading to a decrease in the CO_2 capture efficiency. For the ceramic membrane contactor, when the gas flowrate is 5 L/min, the CO_2 capture efficiency is about 56%, which has not reached the limit of the CO_2 absorption capacity of the solution. Increasing the gas flowrate can improve the CO_2 capture efficiency.

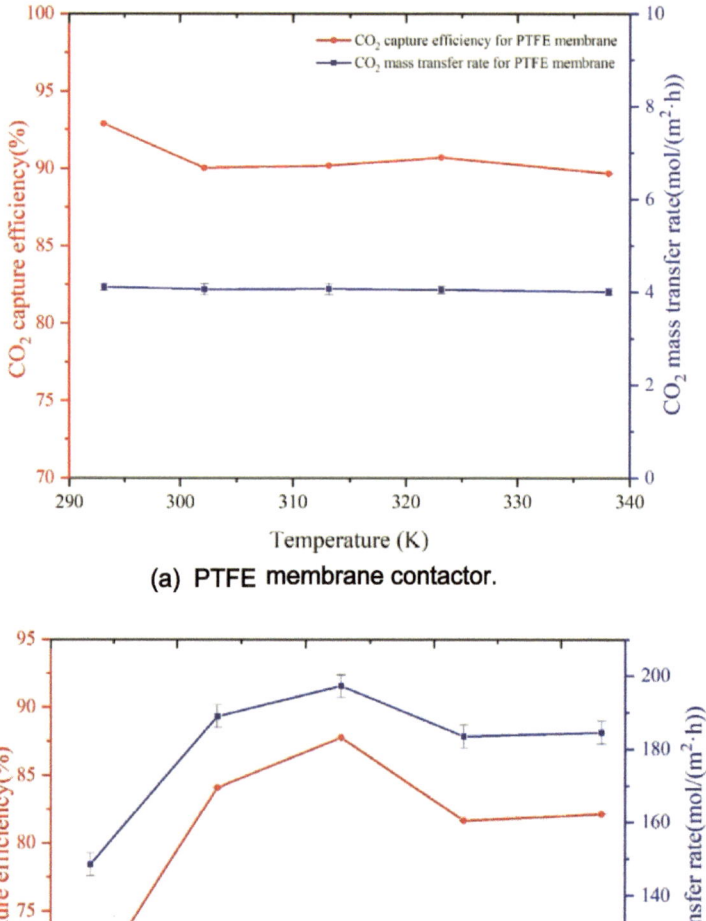

(a) PTFE membrane contactor.

(b) Ceramic membrane contactor.

Fig. 3.11 Effect of the absorber temperature on the capture performance. Reprinted with the permission from Ref. [1] Copyright (2023) (Elsevier)

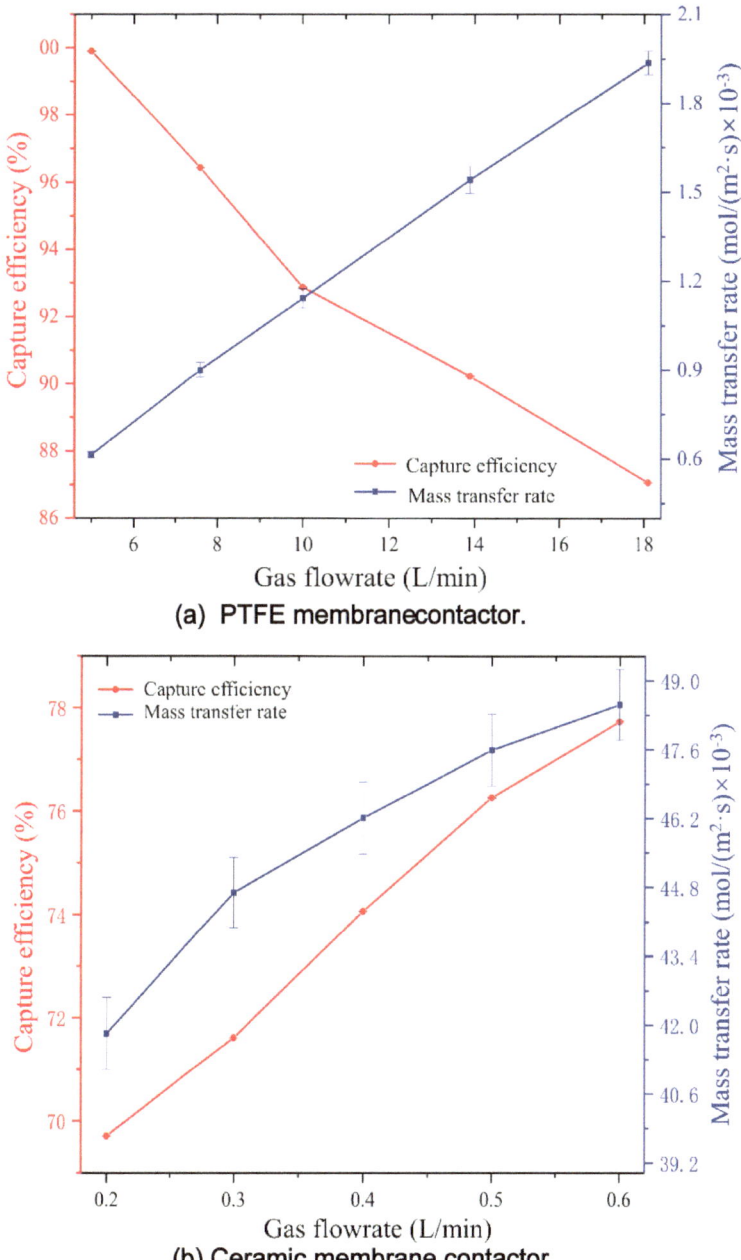

(a) PTFE membranecontactor.

(b) Ceramic membrane contactor.

Fig. 3.12 Effect of the gas flowrate on the capture performance

3.4.4 Effect of the Gas Pressure

Figure 3.13 shows the effect of the gas pressure. When the gas pressure increases, the CO$_2$ capture efficiency and the CO$_2$ mass transfer rate of the two membrane contactors increase. As the gas pressure increases, the gas absorption pressure difference increases. The absorbent pressure remains unchanged, which effectively prevents the membrane wetting and is conducive to the gas mass transfer. Due to the hydrophobicity of the PTFE membrane, the breakthrough pressure is high, and the PTFE membrane is not easy to be wetted. During the experiment, the PTFE membrane contactor avoids wetting even though the gas pressure is lower than the absorbent pressure. For the ceramic membrane contactor, the gas pressure needs to be slightly greater than the absorber pressure to avoid wetting the membrane. Too high gas pressure may cause the absorber circuit in the membrane to bubble or even interrupt the circulation. Therefore, choosing the right gas absorption pressure difference is crucial to improve the CO$_2$ capture efficiency and avoid the membrane wetting.

3.4.5 Comparative Analysis of Experimental Results

According to the experimental results, the CO$_2$ capture efficiency of the PTFE membrane contactor is high, and the CO$_2$ mass transfer rate of the ceramic membrane contactor is high. In experiments, CO$_2$ capture efficiencies of ceramic membrane contactors are about 80%. When the gas flowrate is 10 L/min, the absorbent flowrate is 0.2 L/min, and the membrane pressure difference is 0.4 bar, the CO$_2$ capture efficiency of the ceramic membrane contactor can reach about 90%. It means that under the appropriate operating parameters, ceramic membranes can be used in the CO$_2$ capture industry. The corrosion resistance and the high temperature resistance of organic polymer membrane materials are poor. Ceramic membranes have higher mechanical strength, corrosion and temperature resistances, are easier to install and maintain, and have a longer service life in harsh industrial environments. Therefore, ceramic membranes have more advantages than PTFE membranes in capturing CO$_2$ in the actual flue gas environment.

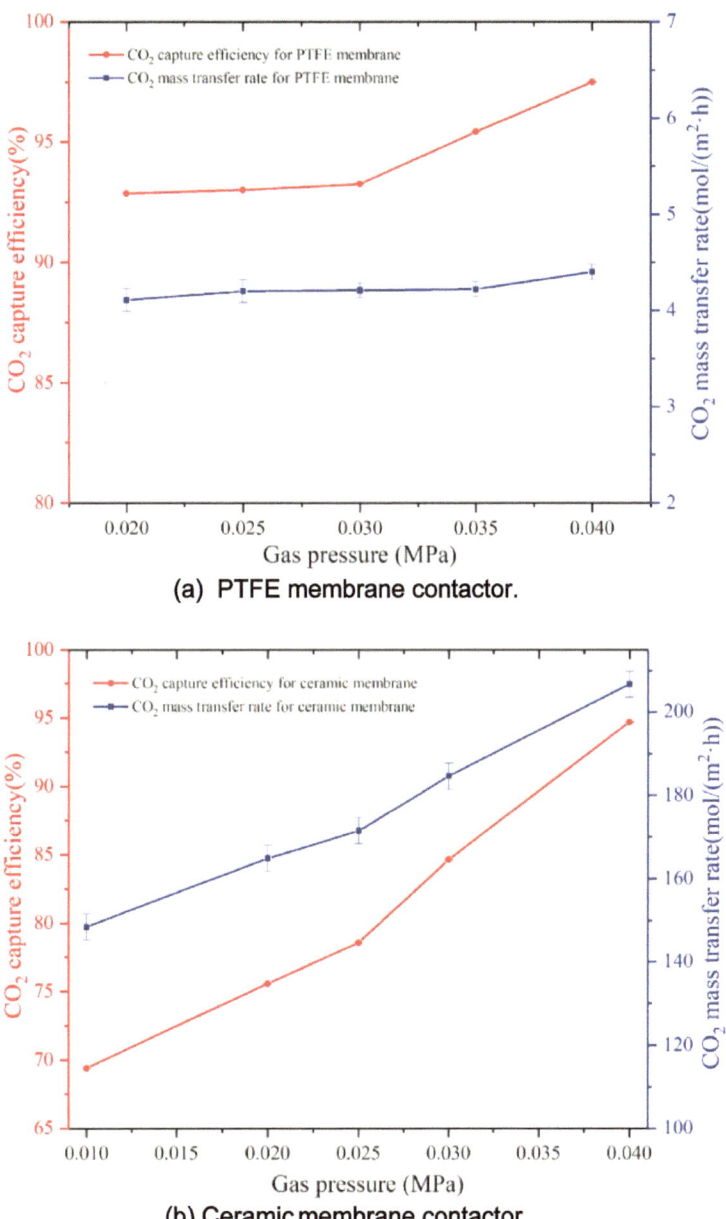

(a) PTFE membrane contactor.

(b) Ceramic membrane contactor.

Fig. 3.13 Effect of the gas pressure on the capture performance. Reprinted with the permission from Ref. [1] Copyright (2023) (Elsevier)

References

1. Fu HM, Xue KL, Li ZH et al (2023) Study on the performance of CO$_2$ capture from flue gas with ceramic and PTFE membrane contactors. Energy 263:125677
2. Yu XH, An L, Yang J et al (2015) CO$_2$ capture using a superhydrophobic ceramic membrane contactor. J Membr Sci 496:1–12
3. Liang DH, Huang JG, Zhang H et al (2021) Influencing factors on the performance of tubular ceramic membrane supports prepared by extrusion. Ceram Int 47:10464–10477
4. Zhang PB, Xu RL, Li HP et al (2019) Mass transfer performance for CO$_2$ absorption into aqueous blended DMEA/MEA solution with optimized molar ratio in a hollow fiber membrane contactor. Sep Purif Technol 211:628–636

Chapter 4
Hydrophobic Modification of Al_2O_3 Ceramic Membrane and Its Application to CO_2 Capture

In order to further improve the CO_2 capturing efficiency and fluxes of the ceramic membrane, the micron alumina ceramic membrane with a larger pore size is selected in this chapter. After the hydrophobic modification, the structure and the mass transfer characteristics of the modified ceramic membrane are studied. The carbon capture performance is verified by experiments, and the crystal phase composition, the surface morphology, the contact angle, the pore size and the porosity of modified ceramic membranes are characterized in details. Furthermore, the operational factors affecting the CO_2 capture efficiency of the hydrophobic Al_2O_3 ceramic membrane are analyzed, and the stability of the modified ceramic membrane is discussed in details.

4.1 Preparation of Hydrophobic Al_2O_3 Ceramic Membrane

The modification process of hydrophobic Al_2O_3 ceramic membranes are shown in Fig. 4.1. Using the impregnation method for the surface modification, the Al_2O_3 ceramic membrane needs to be pretreated first, and then the surface modification material is grafted on it. The main preparation process is as follows.

(1) Ceramic membrane pretreatment: The Al_2O_3 ceramic membrane is ultrasonic cleaned for 30 min with the deionized water, the acetone and the ethanol respectively. Then it is dried at 80 °C for 12 h to increase the concentration of the hydroxyl group (–OH) on the surface of the ceramic membrane.

(2) Preparation of the modified solution: A certain amount of FC16 is mixed with the anhydrous ethanol to prepare 0.1 mol/L graft solution, and then the graft solution is magnetically stirred at room temperature for 12 h to make the graft substance in the solution fully react.

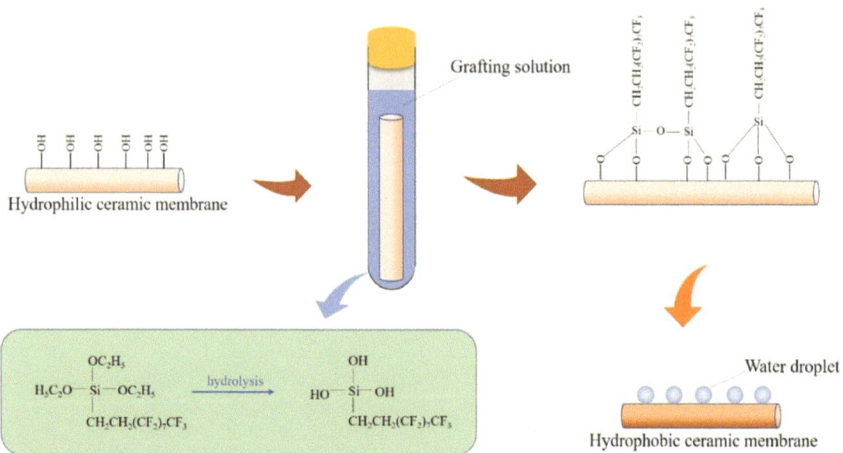

Fig. 4.1 Process of hydrophobic modification of Al$_2$O$_3$ ceramic membrane. Reprinted with the permission from Ref. [1] Copyright (2023) (Elsevier)

(3) Dip coating: The pre-treated Al$_2$O$_3$ ceramic membrane is immersed in the grafting solution at room temperature and left for 12 h. Finally, removing the ceramic membrane, washing it with the deionized water, and drying it at 80 °C for 2 h to obtain the hydrophobic Al$_2$O$_3$ ceramic membrane, as shown in Fig. 4.2. It can be seen that water droplets hover on the surface without penetrating into the membrane, and the ceramic membrane has been successfully modified from hydrophilic to hydrophobic.

During the preparation of the hydrophobic Al$_2$O$_3$ ceramic membrane, the hydroxyl concentration on the surface of the pretreated Al$_2$O$_3$ ceramic membrane increased, and the reaction chance of surfaces of ceramic membranes with the graft solution increased. In addition, the ethoxy group in the activated FC16 solution is replaced by the hydroxyl group to form a new compound. Then, the dehydration

Fig. 4.2 Al$_2$O$_3$ ceramic membrane after hydrophobic modification. Reprinted with the permission from Ref. [1] Copyright (2023) (Elsevier)

condensation reaction with –OH occurs and it is fixed on the surface of the ceramic membrane. At the same time, the grafts also condense with each other and bind together through Si–O–Si bond, thus forming a uniform and stable hydrophobic molecular layer on the surface of the ceramic membrane.

4.2 Experiments on CO_2 Capture Performance of Al_2O_3 Ceramic Membrane

In order to explore effects of Al_2O_3 ceramic membrane on the CO_2 capture performance before and after modification, laboratory-scale CO_2 capture performance experiments are conducted in this chapter. The experimental system is shown in Fig. 4.3, and the parameters of the ceramic membrane contactor are shown in Table 4.1. The whole experimental system is based on a single tube Al_2O_3 ceramic membrane contactor, which is divided into three parts: The flue gas side, the absorber side and the membrane side. The simulated flue gas and the MEA solution flow countercurrent on both sides of the membrane. On the flue gas side, the CO_2 is selectively absorbed by the MEA solution through the transmembrane transport, and the remaining gas is transported to the NaOH solution through pipes and finally discharged to the atmosphere. On the absorbent side, the MEA solution is circulated between the membrane contactor and the absorber tank by a circulating pump, and the absorbent flowrate is measured by a flowmeter. During whole experiments, the pressure difference between the gas and the absorber on both sides of the membrane is maintained at 0.1 bar, and the pressure on the absorber side is higher than that on the flue gas side. Table 4.2 lists the range of experimental parameters. In the process of absorbing CO_2, the Al_2O_3 ceramic membrane cannot participate in the reaction. It simply acts as a barrier for the reaction of the MEA with CO_2, allowing the gas and absorbent to flow on both sides of the membrane, respectively. During the operation of the system, temperatures and pressures at the inlet and outlet are recorded respectively. In addition, each group of experiments is repeated three times, and the average value is taken as the experimental result.

4.3 Evaluation of CO_2 Capture Performance of Al_2O_3 Ceramic Membrane

The CO_2 capture performance of ceramic membranes is mainly evaluated by two indexes: CO_2 capture efficiency and mass transfer rate, which can calculated by Eqs. (3.1) and (3.2) in Chap. 3.

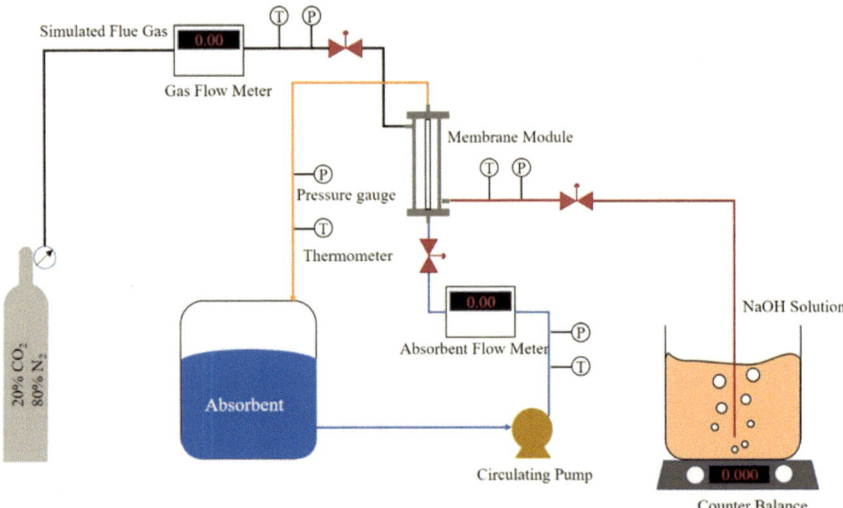

Fig. 4.3 CO$_2$ capture experimental system with Al$_2$O$_3$ ceramic membrane contactor. Reprinted with the permission from Ref. [1] Copyright (2023) (Elsevier)

Table 4.1 Parameters of ceramic membrane contactor

Items	Values
Inner diameter of the ceramic membrane	8 mm
Outer diameter of the ceramic membrane	12 mm
Length of the ceramic membrane	800 mm
Average pore size	0.364 μm (original membrane); 0.358 μm (hydrophobic membrane)
Porosity	43.0% (original membrane); 42.4% (hydrophobic membrane)
Inner diameter of the membrane contactor	30 mm
Outer diameter of the membrane contactor	40 mm
Number of the ceramic membrane	1
Membrane area	0.02 m^2

Table 4.2 Operating parameters of membrane module test for CO$_2$ absorption

Fluids	Parameters	Values
Absorbent	Concentration	5 wt%
	Temperature	25 °C
	Flowrate	0.2–0.6 L/min
Simulated flue gas	CO$_2$ concentration	20%
	Temperature	25 °C
	Flowrate	5–15 L/min

4.4 Analysis of CO$_2$ Capture Application

According to the experimental results, the main crystalline phase of the original ceramic membrane is the corundum phase composed of Al$_2$O$_3$. After hydrophobic modification, the crystalline phase of the ceramic membrane basically cannot change. Therefore, the hydrophobic modification technology has little effect on the crystal phase structure of the ceramic membrane.

Figure 4.4 shows the surface and cross section topography scanning of Al$_2$O$_3$ ceramic membrane before and after modification. After modification, the surface morphology of the ceramic membrane cannot change significantly, and the surface morphology remains basically unchanged. Compared with the original ceramic membrane, the cross section of the modified ceramic membrane is not blocked. In addition, there are no obvious impurities on the surface of the modified ceramic membrane, and the membrane pores are not blocked. It is because the modification process only grafts FC16 molecules to the surface of the ceramic membrane at the molecular level, and the effect on the surface of the membrane is almost negligible. The gas transport resistance of the modified ceramic membrane changes little, so that the gas flux can be kept at a high level.

The contact angle between the pure water and the MEA solution on the Al$_2$O$_3$ ceramic membrane surface before and after modification is shown in Fig. 4.5. The results show that the average contact angle is 47.5° when water droplets start to

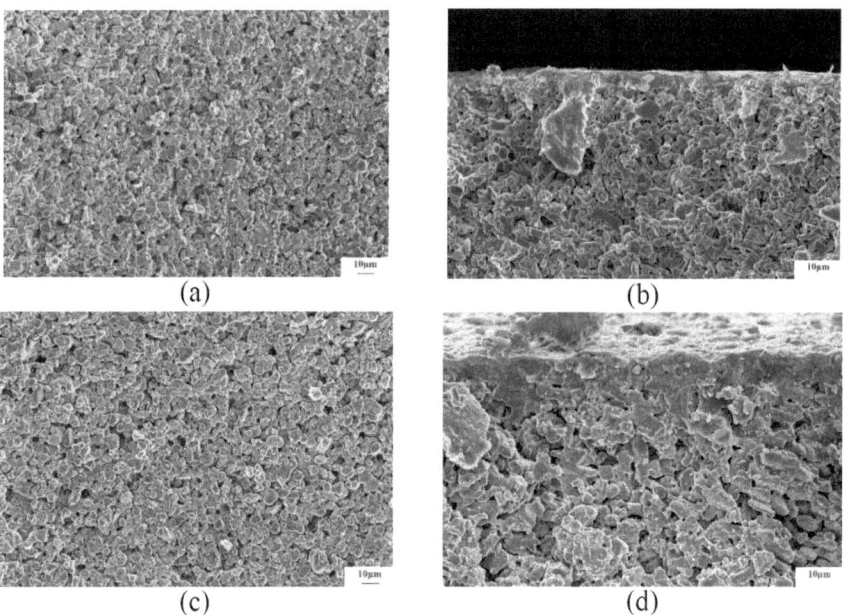

(a) (b) (c) (d)

Fig. 4.4 Surface and cross section morphology of Al$_2$O$_3$ ceramic membrane before (**a** and **b**) and after (**c** and **d**) modification. Reprinted with the permission from Ref. [1] Copyright (2023) (Elsevier)

contact the membrane surface, and it becomes 130.9° when droplets contact the modification membrane surface, which is 163% higher than that of the initial ceramic membrane. As the surface tension of the MEA solution is slightly lower than that of the pure water, contact angles of the MEA solution are slightly smaller than that of the pure water before and after modification, which are about 38° and 122° respectively. In addition, due to the porous structure and the hydrophilicity of the initial ceramic membrane, water droplets begin to penetrate into the membrane as soon as they touch the membrane surface, as shown in Fig. 4.6. The pure water contact angle of the initial ceramic membrane rapidly decreases from the initial 47.5°–6.8° in less than 0.6 s. It can be predicted that the water contact angle on the membrane surface will decrease to 0° within 1 s, and eventually water droplets will completely penetrate into the membrane. On the contrary, after the hydrophobic modification, the contact angle is significantly increased, and the wettability is improved. Water droplets always stay on the surface of the hydrophobic ceramic membrane and cannot penetrate. It can be seen that the modified ceramic membrane exhibits good hydrophobic properties.

In order to investigate the effect of the hydrophobic modification on the mass transfer performance of ceramic membranes, the pore size distribution, N_2 and water fluxes of ceramic membranes before and after modification are measured. Figure 4.7 shows the pore size distribution of ceramic membranes before and after modification. Compared with the original membrane, the pore size of the modified hydrophobic membrane is slightly reduced, and the pore size distribution near the large pore is less. Table 4.3 shows the detailed changes of the pore diameter before and after modification. The average pore diameter of the modified ceramic membrane decreases from 0.364 to 0.358 μm, which is 1.6% less than that before modification.

Figure 4.8 shows the change of N_2 flux of the ceramic membrane with the pressure before and after modification. When the pressure increases, the N_2 flux of the ceramic membrane increases gradually under both conditions. The difference is that the N_2 flux of the modified membrane is slightly smaller than that of the original membrane. Moreover, with the increase of the pressure, the N_2 flux differences increases gradually. Under other conditions, the mass transfer of N_2 through ceramic membrane is greatly affected by the pore size and the porosity [2]. Because the pore size of the membrane is much larger than the mean free path of N_2 molecules, the pore size has less obstruction to N_2 molecules. Therefore, the pore size has little effect on N_2 passing through the membrane. As a result, the porosity is the main factor affecting N_2 flux. Although the average pore sizes of these two kinds of ceramic membranes are not much different, the porosity decreases and the mass transfer resistance increases after modification. Therefore, the N_2 flux of the modified ceramic membrane is lower.

Figure 4.9 shows pure water fluxes of ceramic membranes with pressures before and after modification. Due to the natural hydrophilicity, when the pressure in the initial membrane is slightly increased, water slowly leaks from the membrane. For the modified membrane, when the initial pressure is not higher than the liquid break-through pressure of the ceramic membrane, it is difficult for water to penetrate into the membrane, and the water flux is 0. According to the experiments, the theoretical liquid breakthrough pressure of the modified membrane is 1.8 bar. At the beginning, the water flux of the modified membrane remains 0 with the increase of the pressure.

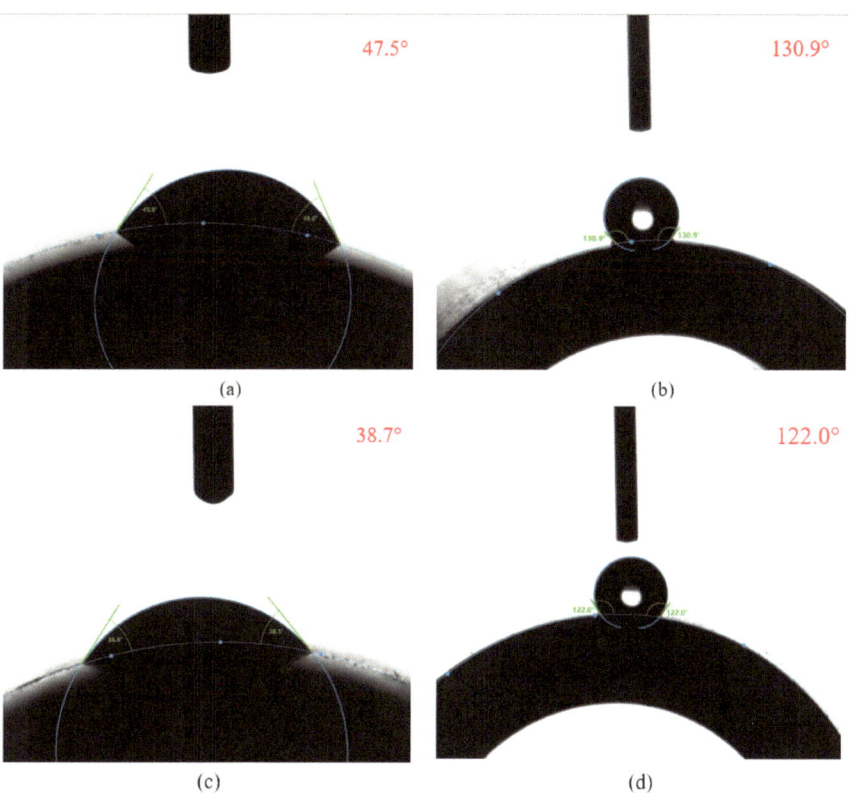

Fig. 4.5 Pure water and MEA solution contact angle of ceramic membrane surface before and after modification (**a** Water contact angle before modification. **b** Water contact angle after modification. **c** MEA solution angle before modification. **d** MEA solution angle after modification). Reprinted with the permission from Ref. [1] Copyright (2023) (Elsevier)

When the pressure increases to 0.5 bar, water begins to seep out of the membrane. At 0.8 bar, water gradually seeps from the membrane surface. When the pressure increases to 1.0 bar, more water seeped out, and the water flux is 0.96 L/(min·m²). The above phenomena indicate that the modified membrane is not easy to be wetted when the pressure difference between inside and outside the membrane is less than 0.5 bar. The actual liquid breakthrough pressure is lower than the theoretical breakthrough pressure, which may be due to the anisotropy of the ceramic membrane and the existence of large pores in the membrane. In addition, compared with the original membrane, the modified membrane is not easily wetted, which is conducive to the mass transfer process of flue gas in the gas–liquid membrane contactor.

The flue gas environment of the thermal power plant is odious, the flue gas temperature is high. In order to determine the thermal stability of the ceramic membrane, the thermogravimetric analysis (TG) is performed on the ceramic membrane before and after the modification, as shown in Fig. 4.10. Due to the thermal evaporation of a

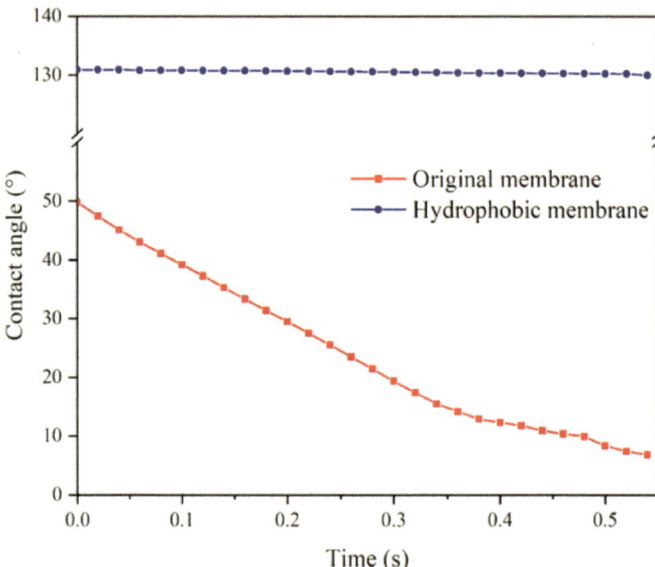

Fig. 4.6 Change of contact angel of ceramic membrane before and after modification. Reprinted with the permission from Ref. [1] Copyright (2023) (Elsevier)

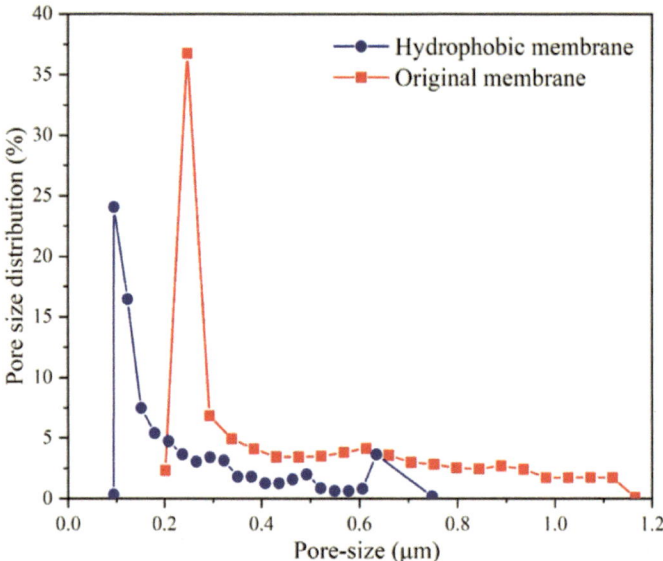

Fig. 4.7 Pore size distribution of ceramic before and after modification. Reprinted with the permission from Ref. [1] Copyright (2023) (Elsevier)

Table 4.3 Pore size of ceramic membrane before and after modification

Membranes	Maximum pore size (μm)	Average pore size (μm)	Most probable pore size (μm)
Original membrane	1.165	0.364	0.247
Hydrophobic membrane	0.783	0.358	0.193

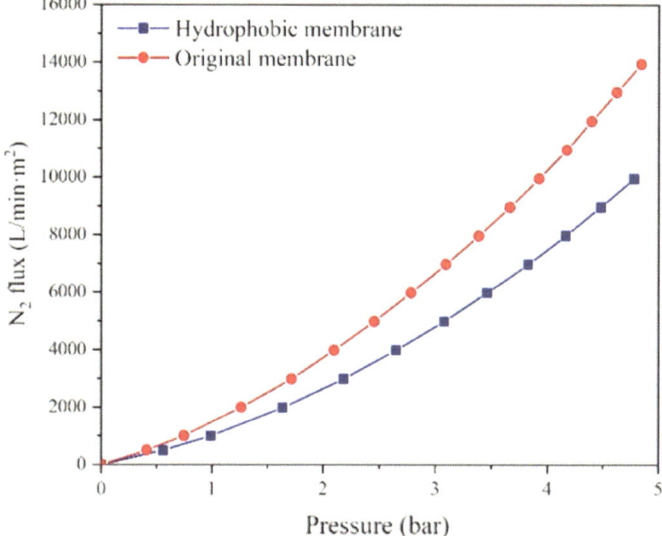

Fig. 4.8 N$_2$ flux of ceramic membrane with pressure before and after modification. Reprinted with the permission from Ref. [1] Copyright (2023) (Elsevier)

small amount of water inside the ceramic membrane, the mass loss of the membrane before and after the modification is greater at 50 °C, and the mass is reduced by 0.55%. After that, as the temperature continues to rise, the quality of the original membrane basically remains stable, and there is no significant weight loss. On the contrary, the mass of hydrophobic membrane decreases from 261.29 to 658.29 °C, and the mass decreases by 1.28%. It is due to the thermal decomposition of FC16 polymers at high temperatures. Qualities of initial and hydrophobic membranes are basically unchanged in the temperature range of 500–800 °C, which reflects that the ceramic membrane has a high temperature resistance. Due to the presence of organosilane molecules on the surface, the hydrophobic membrane can withstand high temperatures of about 260 °C, and its temperature resistance characteristics can meet the temperature requirements of the tail flue of thermal power plants.

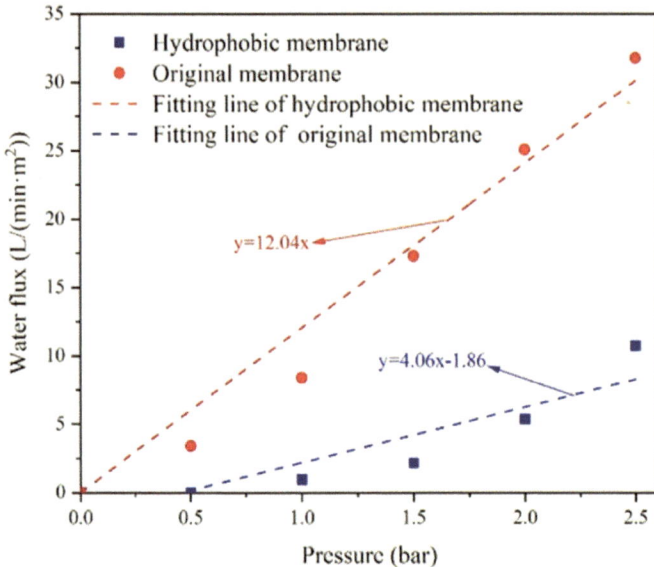

Fig. 4.9 Water flux of ceramic membrane with pressure before and after modification. Reprinted with the permission from Ref. [1] Copyright (2023) (Elsevier)

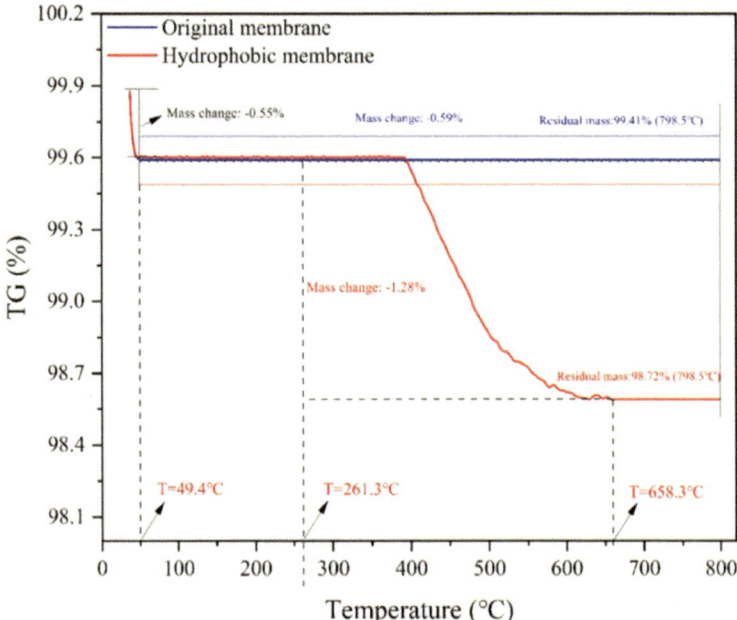

Fig. 4.10 TG and analysis of ceramic membrane before and after modification. Reprinted with the permission from Ref. [1] Copyright (2023) (Elsevier)

4.5 Analysis of CO_2 Capture Performance of Al_2O_3 Ceramic Membrane

In order to research the effect of the ceramic membrane on CO_2 capture performance before and after the modification, effects of the absorbent flowrate and the flue gas flowrate on the CO_2 mass transfer rate and the capture efficiency are experimentally studied, as shown in Fig. 4.11. When the gas flowrate is 5 L/min and the absorber flowrate is 0.6 L/min, the CO_2 mass transfer rate of the hydrophobic ceramic membrane is 18.5×10^{-3} mol/(m^2 s), which is 17.9% higher than that of the initial ceramic membrane. The actual liquid breakthrough pressure of the hydrophobic ceramic membrane is 0.5–0.8 bar. During experiments, the actual pressure difference between the gas and the liquid remains around 0.1 bar, much less than the actual liquid breakthrough pressure of the hydrophobic ceramic membrane. Therefore, the hydrophobic membrane is not easily wetted during the experiment. However, the initial membrane is wetted at the beginning of the experiment when the gas hydraulic difference is 0.1 bar, resulting in the increased mass transfer resistance, and the mass transfer performance of CO_2 is much lower than that of the modified hydrophobic membrane. With the increase of the absorbent flowrate, the CO_2 mass transfer rate increases slowly before and after the modification. It is because the mass transfer resistance mainly comes from the transmembrane transport of CO_2 in the membrane contactor and the diffusion process of CO_2 in the absorber. When the absorbent flowrate increases, the absorbent at the membrane interface transfers rapidly after absorbing CO_2, thus reducing the concentration polarization phenomenon at the gas–liquid interface. In addition, with the increase of the absorbent flowrate, the fluid boundary layer thickness at the absorbent side interface decreases gradually. The mass transfer resistance is further reduced, so the mass transfer rate of CO_2 increases with the increase of the absorbent flowrate. It is worth noting that with the increase of the flowrate, the curve of CO_2 mass transfer rate gradually flattens. It is due to the increase in the CO_2 absorption flux as the absorbent flowrate increases. Figure 4.12 shows the influence of the absorbent flowrate on CO_2 capture efficiencies of ceramic membranes before and after the modification, and trends of curves are consistent with mass transfer rates. With the increase of the absorbent flowrate, the CO_2 capture efficiency of the modified ceramic membrane increases from 87.3 to 98.0%, while the CO_2 capture efficiency of the original ceramic membrane only increases from 73.0 to 80.2%. Therefore, the absorbent flowrate has a significant positive effect on the CO_2 capture performance of the modified membrane. In practical engineering applications, the CO_2 mass transfer rate and the CO_2 capture efficiency can be improved by increasing the absorbent flowrate.

As shown in Fig. 4.13, when the absorbant flowrate is 0.2 L/min and the flue gas flowrate gradually increases from 5 to 15 L/min, the CO_2 mass transfer rate of ceramic membranes before and after the modification almost increase linearly. When the flue gas flowrate is 15 L/min, the CO_2 mass transfer rate of the hydrophobic ceramic membrane reaches 46.6×10^{-3} mol/(m^2 s), which is 15.7% higher than that of the initial ceramic membrane. Although the CO_2 mass transfer rate increases with the

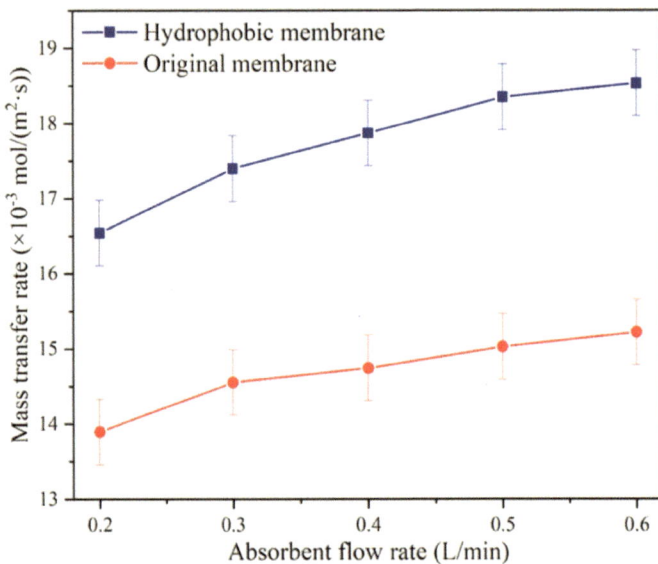

Fig. 4.11 Effect of the absorbent flowrate on the CO_2 mass transfer rate of ceramic membranes before and after the modification (gas flowrate is 5 L/min). Reprinted with the permission from Ref. [1] Copyright (2023) (Elsevier)

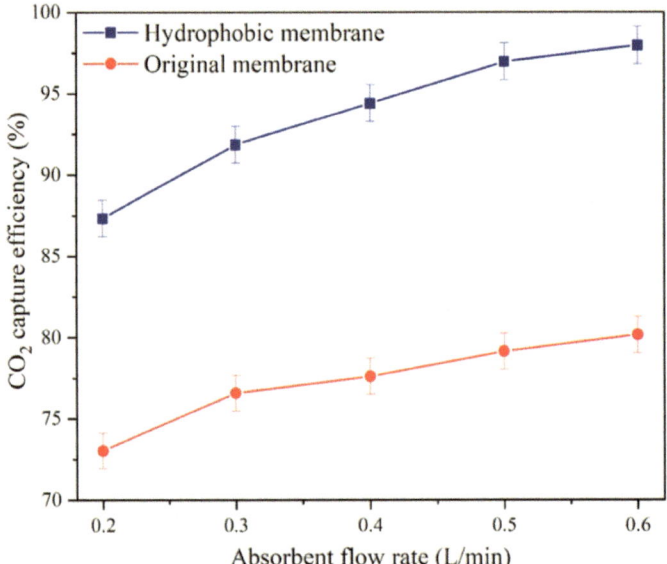

Fig. 4.12 Effect of the absorbent flowrate on the CO_2 capture efficiency of ceramic membranes before and after the modification (gas flowrate is 5 L/min). Reprinted with the permission from Ref. [1] Copyright (2023) (Elsevier)

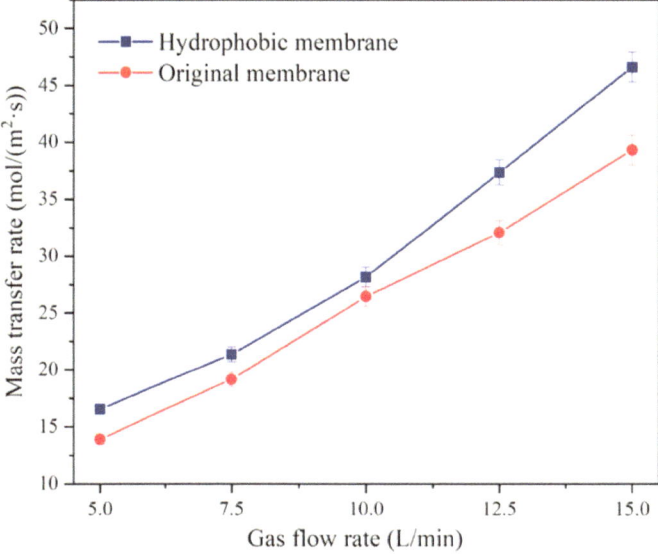

Fig. 4.13 Effect of the gas flowrate on the CO$_2$ mass transfer rate of ceramic membranes before and after the modification (absorbent flowrate is 0.2 L/min). Reprinted with the permission from Ref. [1] Copyright (2023) (Elsevier)

increase of the flue gas flowrate, the residence time of the flue gas in the membrane contactor is shortened. As a result, the contact time between CO$_2$ and the absorber is reduced. Some of the CO$_2$ is expelled from the membrane contactor before it reacts with the absorbent, resulting in reducing the CO$_2$ capture efficiency. Figure 4.14 shows that the CO$_2$ capture efficiency of the modified membrane decreases with the increase of the flue gas flowrate. It is worth noting that the CO$_2$ capture efficiency of the hydrophobic ceramic membrane decreases rapidly at first, and then slowly. Although increasing the flue gas flowrate can increase the CO$_2$ mass transfer rate, the residence time of the flue gas in the membrane contactor is shortened, and the CO$_2$ capture efficiency of the membrane is rapidly reduced.

4.6 Corrosion Resistance of Al$_2$O$_3$ Ceramic Membranes

In membrane contactors, alkaline absorbers are usually utilized. In a long-term alkaline environment, the membrane is easy to be corroded. It results in a sharp decline in the membrane performance, and greatly reduces the service life. Therefore, the immersion method of MEA solution is used to study the corrosion of ceramic membrane by controlling variables. First, ceramic membranes are pretreated and ultrasonic cleaned for 10 min. Second, drying in a vacuum oven for 12 h. Pore size distributions and masses of ceramic membranes are measured after completely

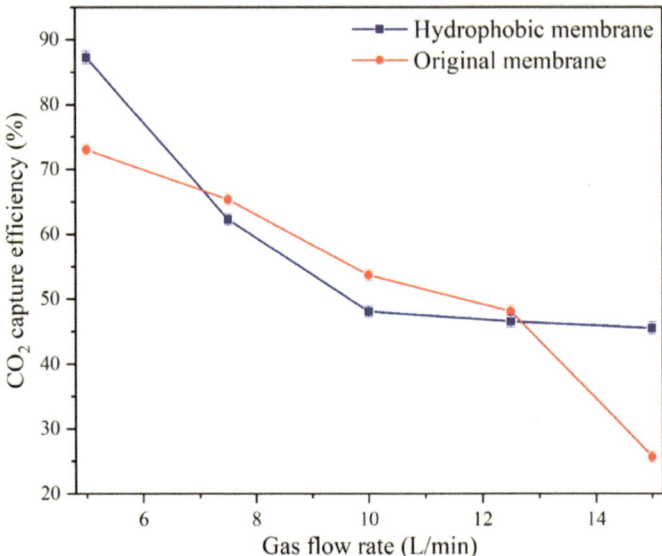

Fig. 4.14 Effect of the gas flowrate on the CO_2 capture efficiency of ceramic membranes before and after the modification (absorbent flowrate is 0.2 L/min). Reprinted with the permission from Ref. [1] Copyright (2023) (Elsevier)

drying. Third, it is soaked in MEA solutions of different concentrations. Fourth, after soaking for a period of time, removing it and ultrasonic cleaning. Finally, the aperture distribution and quality are analyzed. Table 4.4 describes the operation parameters. In the experiment, the mass change rate and the average pore diameter are used as the evaluation indexes of the corrosion resistance. The mass change rate is calculated according to Eq. (4.1), and the average aperture is measured by the aperture analyzer.

$$r_m = \frac{|m_a - m_b|}{m_a} \tag{4.1}$$

Table 4.4 Operating parameters of the corrosion resistance test for ceramic membranes

Soaking time (h)	MEA concentration (wt%)
12	5
24	10
48	20
72	30
/	100

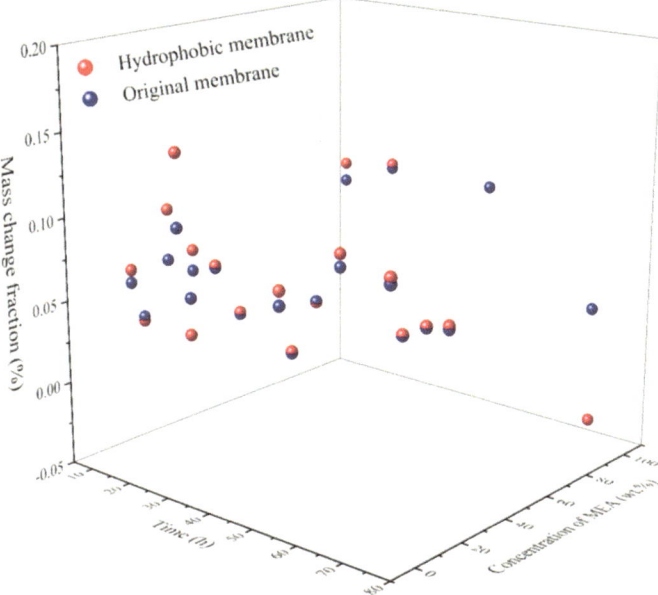

Fig. 4.15 Effect of the immersion time and the MEA concentration on masses of ceramic membranes before and after the modification. Reprinted with the permission from Ref. [1] Copyright (2023) (Elsevier)

where "r_m" represents the mass change rate of the ceramic membrane. "m_a" and "m_b" represent the masses of ceramic membranes before and after the impregnation, respectively.

Figure 4.15 shows the influence of the soaking time and the MEA solution concentration on the mass change rate of ceramic membranes before and after the modification. After soaking in the MEA solution of different concentrations for a period of time, the masses of ceramic membranes change slightly, but the change rates are less than 0.15%. It may be due to some MEA molecules binding to ceramic membranes. Figure 4.16 shows the influence of the soaking time and the MEA concentration on the average pore diameter of ceramic membranes before and after the modification. After the immersion in the MEA solution, average pore diameters of ceramic membranes vary between 0.26 and 0.32 μm. In addition, the average pore size of the hydrophobic membrane is slightly lower than that of the original membrane. There is no significant relationship among the average pore size, the soaking time and the concentration of the MEA solution. The variation of the average pore diameter may be caused by the anisotropy of the ceramic material. Therefore, the long-term immersion in the MEA solution causes little corrosion to the ceramic membrane, and the probability of the pores being blocked is small. Ceramic membranes show good stability in the corrosion resistance.

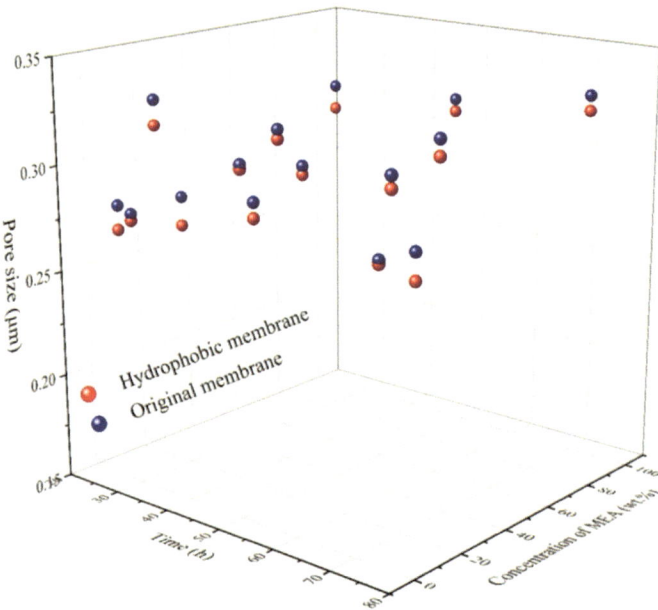

Fig. 4.16 Effect of the immersion time and the MEA concentration on average pore sizes of the hydrophobic and the original ceramic membranes. Reprinted with the permission from Ref. [1] Copyright (2023) (Elsevier)

References

1. Fu HM, Xue KL, Yang JH et al (2023) CO$_2$ capture based on Al$_2$O$_3$ ceramic membrane with hydrophobic modification. J Eur Ceram Soc 43:3427–3436
2. Ngoi HY, Lai LS, Tay WH et al (2022) Computational fluid dynamics simulation for dual-phase membrane module for CO$_2$ capture: effect of membrane thickness, temperature and CO$_2$ feed concentration. Environ Prog Sustain 41:e13780

Chapter 5
Analysis of CO_2 Mass Transfer Performance of Superhydrophobic Ceramic Membranes

The mass transfer coefficient is an important performance evaluation index of membrane contactors, which directly affects the gas flux and the carbon capture performance. The mass transfer coefficient can be improved by various methods, such as using the gas–liquid cross flow [1], using high porosity hydrophobic membranes, adjusting the arrangement of contactor inner membranes and developing new absorbers [2]. In this chapter, based on the sol–gel method and the impregnation method, surface modifications of two kinds of Al_2O_3 ceramic membranes with different membrane thickness are carried out to prepare superhydrophobic ceramic membranes. At the same time, structures and properties of the two kinds of membranes are characterized in detail. Finally, it is applied to experiments of the CO_2 capture, and the influence of the superhydrophobicity on the mass transfer coefficient of the ceramic membrane is discussed.

5.1 Preparation of Superhydrophobic Al_2O_3 Ceramic Membranes

The surface roughness and the surface energy of ceramic membranes are modified by the sol–gel method and the impregnation method. First, ceramic membranes M1 and M2 are ultrasonic cleaned with the anhydrous ethanol, the acetone and the deionized water respectively for 30 min, and they are dried at 80 °C. Second, the ethyl orthosilicate, the methanol and the ammonia are mixed in a certain proportion and left them at room temperature for 12 h to fully react to form the silica sol. Third, a certain amount of 1H,1H,2H,2H-perfluorodecyl triethoxysilane is added to the mixture and is stirred at 60 °C for 24 h. The mass ratio of the ethyl orthosilicate, the methanol, the ammonia and 1H,1H,2H,2H-perfluorodecyl triethoxysilane is 26: 153: 35: 2.16. Fourth, ceramic membranes cleaned by the deionized water and the ethanol are immersed in the mixture at 80 °C for 12 h for grafting the modified colloidal

© The Author(s), under exclusive license to Springer Nature Switzerland AG 2024
Z. Li et al., *Hydrophobic Ceramic Membranes for CO2 Capture*,
SpringerBriefs in Energy, https://doi.org/10.1007/978-3-031-77678-6_5

silica. Finally, ceramic membranes clean again to remove surface impurities and dry again for 12 h.

5.2 Experimental Analysis of CO$_2$ Capture Performance of Ceramic Membranes

The flow diagram and photograph of the experimental system using ceramic membrane contactor to capture CO$_2$ are shown in Figs. 5.1 and 5.2, respectively. The whole experimental platform is divided into two parts: the gas side and the tube side. On the gas side, simulated flue gas enters the ceramic membrane contactor from the stainless steel cylinder through the pressure reducing valve and the flow control valve. It then reacts with the absorbent on the tube side and exits through the membrane contactor gas outlet, where the CO$_2$ concentration is measured by a portable handheld CO$_2$ concentration analyzer. On the tube side, the absorbent enters the tube of the ceramic membrane contactor through the flow meter and the valve under the action of the circulating pump. The absorber on the tube side selectively absorbs CO$_2$ from the flue gas, and then flows from the tube into the enriched liquid tank. Thermocouples and digital pressure gauges are installed at inlets and outlets to track changes in temperatures and pressures, and data from experiments are recorded via paperless recorders. Table 5.1 lists parameters of the ceramic membrane contactor, and Table 5.2 lists the operating parameters during experiments.

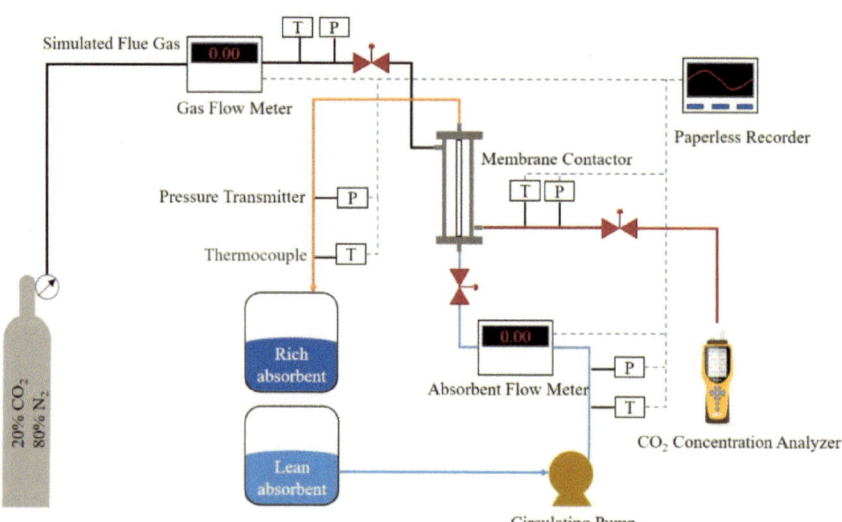

Fig. 5.1 Schematic diagram for CO$_2$ capture experiments by the ceramic membrane contactor. Reprinted with the permission from Ref. [3] Copyright (2023) (Elsevier)

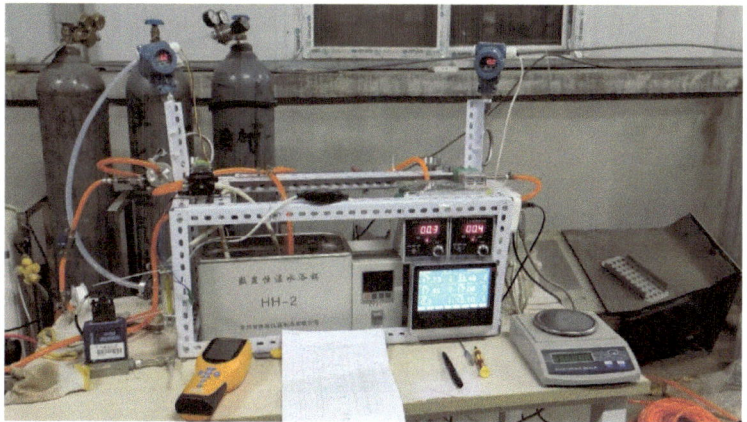

Fig. 5.2 Photograph for CO_2 capture experiments by the ceramic membrane contactor

Table 5.1 Parameters of M1 and M2 ceramic membranes and contactors

Items	M1	M2
Outer diameter of the membrane (mm)	12.5	12
Inner diameter of the membrane (mm)	9.5	8
Average pore size (μm)	0.46	0.49
Porosity	25.7	38.8
Tortuosity	2.4	2.6
Effective length (mm)	800	800
Effective membrane area ($\times 10^{-3}$ m^2)	19.63	18.84
Outer diameter of the module (mm)	40	40
Inner diameter of the module (mm)	30	30
Filling rate	0.17	0.16

Table 5.2 Operation parameters of CO_2 capture experiments

Items	Values
Velocity of the absorbent (m/s)	0.005–0.024
Velocity of the gas (m/s)	0.014–0.069
Temperature (K)	298.15
CO_2 concentration (vol%)	20
MEA concentration (wt%)	20

5.3   Mass Transfer Theory and Performance Evaluation of CO_2 Capture

5.3.1   Evaluation of CO_2 Capture Performance of Ceramic Membranes

The total mass transfer coefficient and the CO_2 capture efficiency on the gas side are used to evaluate the CO_2 capture performance of the modified ceramic membrane, which can be calculated by Eqs. (5.1)–(5.3) [4].

$$J_{CO2} = \frac{Q_{g,in}C_{CO2,in} - Q_{g,out}C_{CO2,out}}{A} = K_g \Delta C_g \tag{5.1}$$

$$\Delta C_g = \frac{C_{CO2,in} - C_{CO2,out}}{In(C_{CO2,in}/C_{CO2,out})} \tag{5.2}$$

$$\eta = \left(1 - C_{CO2,out}/C_{CO2,in}\right) \times 100\% \tag{5.3}$$

where "J_{CO2}" represents the absorbed flux of CO_2. "$Q_{g,out}$" represents the outlet gas flowrate in the membrane contactor. "$C_{CO2,in}$" and "$C_{CO2,out}$" represent the CO_2 concentrations in the inlet and outlet flue gases, respectively. "K_g" represents the total mass transfer coefficient. "ΔC_g" represents the log-average driving force based on the gas phase concentration. "η" represents the CO_2 capture efficiency.

Considering the effect of structure parameters of the membrane contactor on the gas mass transfer performance, dimensionless parameters are used to describe the CO_2 capture performance of ceramic membranes before and after the modification. The Re number is used to indicate the gas and the absorber flowrates in the membrane contactor, and the mass transfer performance of gas is expressed by the Sh number.

$$Re = \rho_f ud/v \tag{5.4}$$

$$Sh = kd/D \tag{5.5}$$

where "ρ_f" represents the density of the fluid. "u" represents the velocity of the fluid. "d" represents the diameter of the membrane contactor. "v" represents the dynamic viscosity. "k" represents the mass transfer coefficient. "D" represents the diffusion coefficient of the fluid.

5.3.2 Analysis of Mass Transfer Resistance in Membranes

The mass transfer process in a ceramic membrane contactor is shown in Fig. 5.3. According to the series resistance model [5], overall mass transfer resistances mainly come from three parts: The gas phase resistance, the membrane phase resistance and the liquid phase resistance. The relationship between total mass transfer resistances and the partial resistance is shown in Eqs. (5.6) and (5.7) [6].

$$\frac{1}{K_g} = \frac{d_i}{k_g} + \frac{d_i}{d_{ln}k_m} + \frac{1}{EH_d k_l} \tag{5.6}$$

$$H_d = RT/H \tag{5.7}$$

where "k_g", "k_m", and "k_l" represent the mass transfer coefficients of the gas, the membrane, and the liquid, respectively. "d_i" and "d_{ln}" represent the inner and the logarithmic mean diameters of the ceramic membrane, respectively. "H_d" represents the dimensionless Henry constant. "R" represents the universal gas constant. "T" represents the temperature. "H" represents the Henry constant. "E" represents the enhancement factor, which can be calculated by Eq. (5.8).

$$E = \frac{-Ha^2}{2(E_\infty - 1)} + \sqrt{\frac{Ha^4}{4(E_\infty - 1)^2} + \frac{E_\infty Ha^2}{E_\infty - 1} + 1} \tag{5.8}$$

where "Ha" represents the Hatta number, which can be calculated by Eq. (5.9). "E_∞" represents the infinite enhancement factor, which can be calculated by Eq. (5.10).

$$Ha = \frac{\sqrt{r_{CO2,MEA} D_{CO2,MEA} C_{MEA}}}{k_l} \tag{5.9}$$

$$E_\infty = \left(1 + \frac{C_{MEA} D_{MEA,l}}{v_{MEA} C_{CO2,i} D_{CO2,MEA}}\right)\left(\frac{D_{CO2,MEA}}{D_{MEA,l}}\right)^{1/3} \tag{5.10}$$

where "$r_{CO2,MEA}$" represents the second order reaction rate constant of CO_2 and the MEA reaction. "$D_{CO2,MEA}$" represents the diffusion coefficient of CO_2 in the MEA solution. "C_{MEA}" represents the concentration of the MEA solution. "$D_{MEA,l}$" represents the diffusion coefficient of the MEA solution in the liquid. "v_{MEA}" represents the stoichiometric coefficient of MEA. "$C_{CO2,i}$" represents the concentration of CO_2 at the liquid phase interface, which can be calculated by Eq. (5.11).

$$C_{CO2,i} = \left[\frac{p_{CO2} + (k_l d_i E C_{CO2,L})/(d_o k_{gt})}{1 + (k_l d_i E)/(d_o k_{gt})}\right] H \tag{5.11}$$

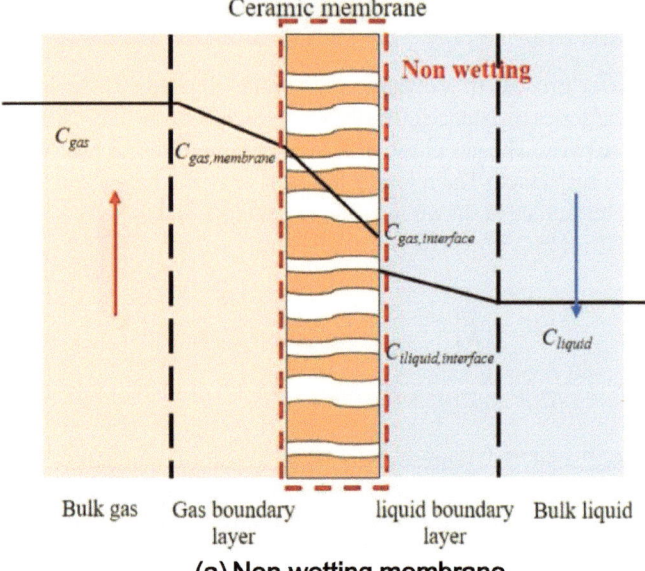

(a) Non-wetting membrane

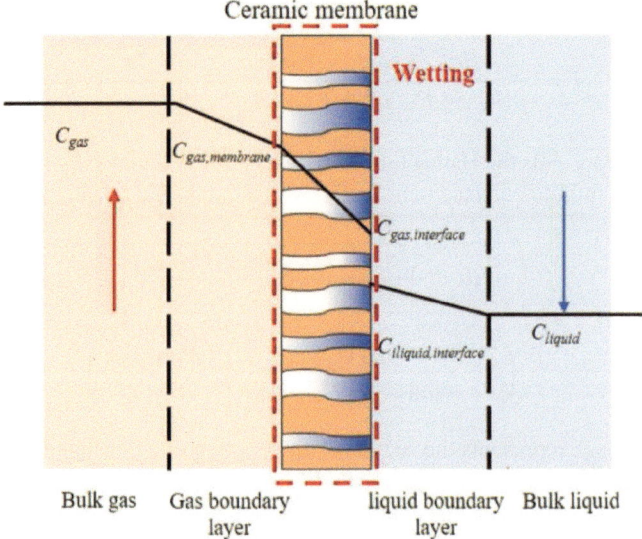

(b) Partial -wetting membrane

Fig. 5.3 Mass transfer process of the non-wetting membrane and the partial-wetting membrane. Reprinted with the permission from Ref. [3] Copyright (2023) (Elsevier)

where "p_{CO2}" represents the partial pressure of CO_2 in the flue gas. "$C_{CO2,L}$" represents the concentration of CO_2 in the liquid phase. "k_{gt}" represents total mass transfer coefficients of the gas and the liquid phases, which can be calculated by Eq. (5.12).

$$k_{gt} = \frac{k_g/(RT)}{1 + k_g/k_m}. \qquad (5.12)$$

When the flue gas flows on the membrane contactor shell side, the relationship between the gas mass transfer coefficient, Re number and Sc number is shown as Eqs. (5.13) and (5.14). It should be noted that these equations are available for Re number is in the range of 0–500, and the filling factor is in the range of 0.04–0.4.

$$Sh = 5.8(1 - \varphi)(d_o/L)Re^{0.6}Sc^{0.3} \qquad (5.13)$$

$$Sc = \nu/D_{CO2-gas} \qquad (5.14)$$

where "φ" represents the filling factor. "d_o" and "L" represent the outer diameter and the length of the ceramic membrane, respectively. "$D_{CO2-gas}$" represents the diffusion coefficient of CO_2 in the gas phase.

For non-wetting ceramic membranes, the mass transfer coefficient of the membrane can be calculated by Eq. (5.15).

$$k_m = (D_{CO2,m}\varepsilon)/(\delta\tau_m) \qquad (5.15)$$

where "$D_{CO2,m}$" represents the diffusion coefficient of CO_2 in the membrane. "ε", "δ", and "τ_m" represent the porosity, the thickness, and the tortuosity of the ceramic membrane, respectively.

Since the mass transfer process of CO_2 in the ceramic membrane includes the molecular diffusion and the Knudsen diffusion, the mass transfer coefficient of CO_2 in the ceramic membrane can be expressed as Eq. (5.16) [7].

$$\frac{1}{D_{CO2,m}} = \frac{1}{D_M} + \frac{1}{D_k} \qquad (5.16)$$

where "D_M" represents the molecular diffusion coefficient, which can be calculated by Eq. (5.17). "D_k" represents the Knudsen diffusion coefficient, which can be calculated by Eq. (5.18).

$$D_M = \frac{1200\nu RT\Omega_\nu}{M_{CO2}p\Omega_d} \qquad (5.17)$$

$$D_k = \frac{2r}{3}\sqrt{\frac{8RT}{\pi M_{CO2}}} \qquad (5.18)$$

where "M_{CO2}" represents the relative molecular mass of CO_2. "p" represents the pressure. "r" represents the pore size of the membrane. "Ω_v" and "Ω_d" represent the viscous and the diffusion collision integrals, respectively, which can be calculated by Eqs. (5.19) and (5.20).

$$\Omega_v = \frac{1.16145}{T_d^{0.14675}} + \frac{0.52487}{\exp^{0.7732T_d}} + \frac{2.16178}{\exp^{2.43787T_d}} \tag{5.19}$$

$$\Omega_d = \frac{1.6036}{T_d^{0.1561}} + \frac{1.6036}{\exp^{0.47635T_d}} + \frac{1.033587}{\exp^{1.52996T_d}} + \frac{1.76476}{\exp^{3.89411T_d}} \tag{5.20}$$

where "T_d" represents the dimensionless temperature, which can be calculated by Eq. (5.21).

$$T_d = \kappa T / \lambda \tag{5.21}$$

where "κ" represents the Boltzmann constant. "λ" represents the Stockmayer constant.

The mass transfer coefficient of the liquid phase depends on the flowrate of the absorber and the diffusion coefficient of CO_2 in the MEA solution, which can be calculated by Eq. (5.22).

$$k_l = \frac{D_{CO2,MEA}}{d_i} \left[(3.47)^3 + (1.315)^3 \frac{d_i^2 v_i}{D_{CO2,MEA}L} \right] \tag{5.22}$$

where "$D_{CO2,MEA}$" represents the diffusion coefficient of CO_2 in the MEA solution. "v_i" represents the absorbent velocity.

5.4   Characterization Results of Ceramic Membranes

Figure 5.4 shows the surface and the cross section morphologies of M1 and M2 before and after the modification. It can be seen that the surface and the cross sections (a, c, e and g) of M1 and M2 membranes are composed of a large number of massive micron Al_2O_3 particles stacked, showing a typical asymmetric structure. After the hydrophobic modification, SiO_2 particles are attached to the surface of M1 and M2 (red box part in Fig. 5.4), and the particle size of SiO_2 particles is between 0.5–1 μm. Therefore, when SiO_2 particles are fixed on the surface of the membrane, the surface roughness of the membrane is easily changed. However, no SiO_2 particles are observed on the cross sections of M1 and M2, indicating that the modification process only changes the structure of the surface, and cannot clog the internal pores of M1 and M2. By observing the surface morphology of the modified M1 and M2, it is confirmed that SiO_2 particles have been successfully fixed on the surface of M2 and M2 after the sol–gel and the impregnation processes.

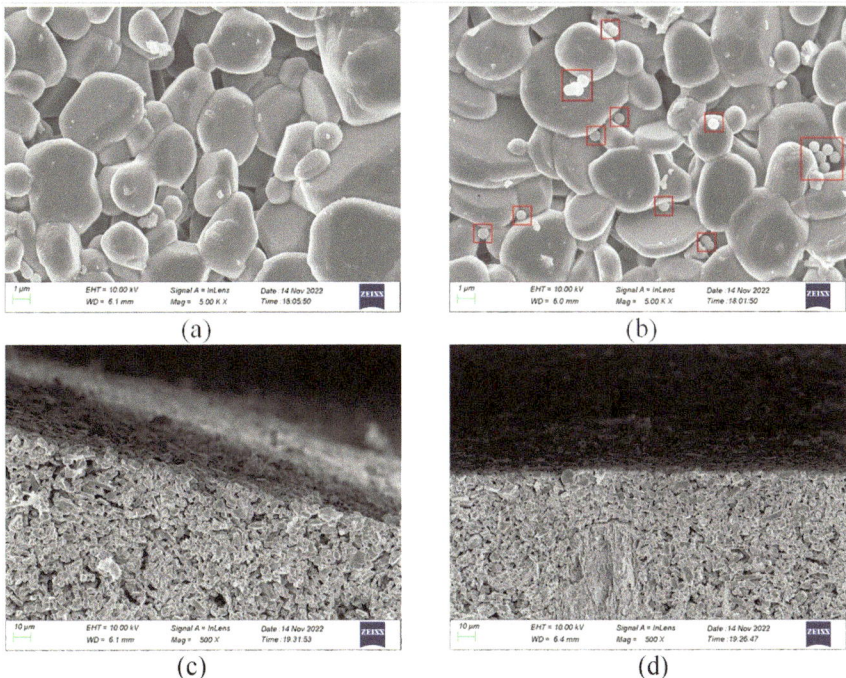

Fig. 5.4 Morphology of M1 and M2 before and after the modification (**a** surface of M1 before the modification; **b** surface of M1 after the modification; **c** cross section of M1 before the modification; **d** cross section of M1 after the modification; **e** surface of M2 before the modification; **f** surface of M2 after the modification; **g** cross section of M2 before the modification; **h** cross section of M2 after the modification). Reprinted with the permission from Ref. [3] Copyright (2023) (Elsevier)

Figure 5.5 shows three-dimensional surface contours and corresponding surface roughness parameters of M1 and M2 before and after the modification. Before the modification, surfaces of M1 and M2 are relatively flat, there is no obvious peak, and the surface roughness distribution is relatively uniform. In contrast, cluster structures can be observed on the surfaces of the modified M1 and M2, and the surface mean roughness (R_a) and the root mean square roughness (R_q) are significantly increased. It is caused by the accumulation of SiO_2 particles on surfaces of M1 and M2. As a result, SiO_2 particles have been successfully fixed on surfaces of M1 and M2, and the roughness of M1 and M2 surfaces increases after the impregnation.

In order to verify the chemical composition of surfaces, the surface composition of M1 and M2 before and after the modification is analyzed by the Fourier infrared spectroscopy. After the modification, there are absorption peaks on surfaces of M1 and M2 at 1200 and 1144 cm^{-1}, which is caused by the C-F tensile vibration of the functional groups $–CF_2$ and $–CF_3$, which provides a basis for the successful grafting of fluorosiloxane on the membrane surface. The absorption peaks at 1090 and 461 cm^{-1} correspond to SiO_2 particles and Si–O–Si vibrations of SiO_2 produced by the hydrolytic polycondensation of the ethyl orthosilicate [8]. In addition, samples

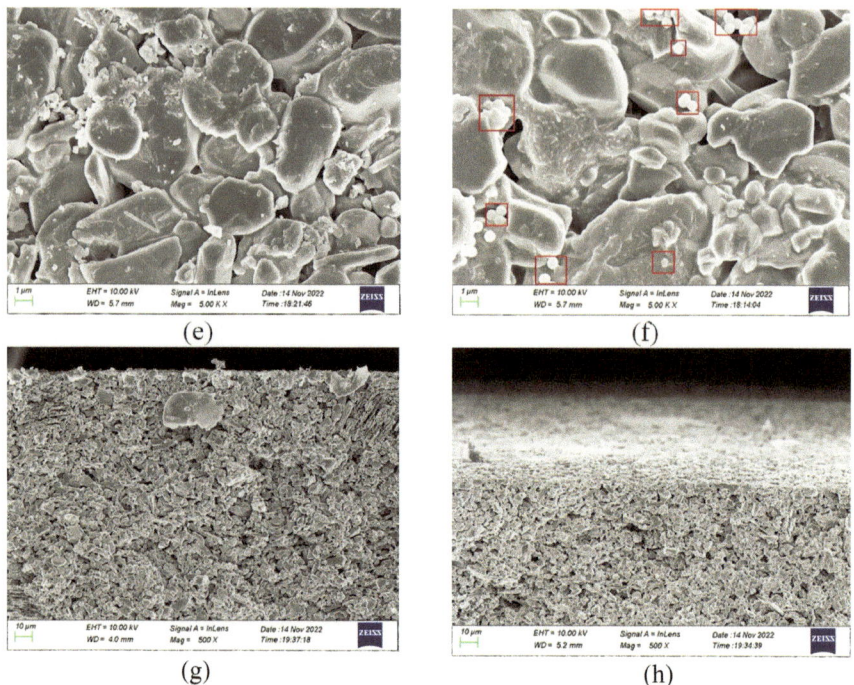

Fig. 5.4 (continued)

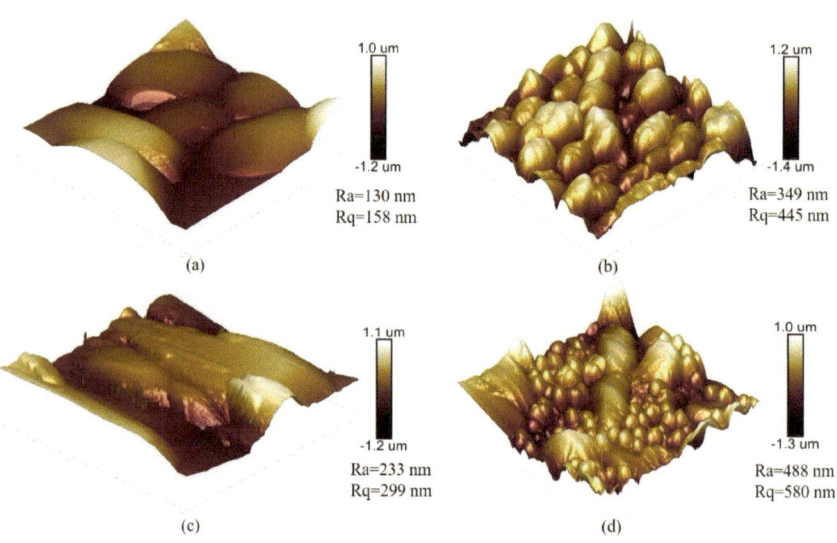

Fig. 5.5 Three-dimensional surface profiles of M1 and M2 before and after the modification (**a** M1 before the modification; **b** M1 after the modification; **c** M2 before the modification; **d** M2 after the modification). Reprinted with the permission from Ref. [3] Copyright (2023) (Elsevier)

show wide absorption peaks at 3400–3600 cm^{-1} and 1634 cm^{-1}, which are due to the stretching and bending vibrations of hydroxyl (–OH) and silica groups (Si–OH). Through the above analysis, it can be determined that the SiO_2 particles and fluorosilane molecules have been fixed on the surface of the ceramic membrane.

In order to study the effect of the hydrophobic modification on the pore size and other structural parameters of ceramic membranes, mean pore sizes, porosities and N_2 fluxes of M1 and M2 are measured before and after the modification. Figure 5.6 shows average pore sizes of M1 and M2 before and after the modification. Average pore sizes of both membranes before the modification are not much different, while average pore sizes of both membranes after the modification are slightly decreased. The M1 pore size decreases from 0.46 to 0.45 μm, while the M2 pore size decreases from 0.49 to 0.44 μm. Figure 5.7 shows the changes of M1 and M2 porosities before and after the modification. The porosity of M1 is about 10% smaller than that of M2. After the modification, porosities of M1 and M2 decrease slightly. After the modification, SiO_2 particles cannot enter the ceramic membrane, because the effect of SiO_2 particles plugging in the ceramic membrane on the pores is negligible.

In order to further verify the gas mass transfer performance of ceramic membranes before and after the modification under dry conditions, N_2 fluxes of M1 and M2 before and after the modification are tested (Fig. 5.8). Through linear fitting of experimental data, slopes of N_2 fluxes of M1 and M2 before and after the modification are obtained, namely mass transfer coefficients of membranes. N_2 mass transfer coefficients of M1 and M2 membranes decrease slightly after the modification. The mass transfer coefficient of M1 decreases from 72.88 L/(m^2 min kPa) to 61.78 L/(m^2 min kPa), and the mass transfer coefficient of M2 decreases from 22.55 L/(m^2 min kPa) to 16.72

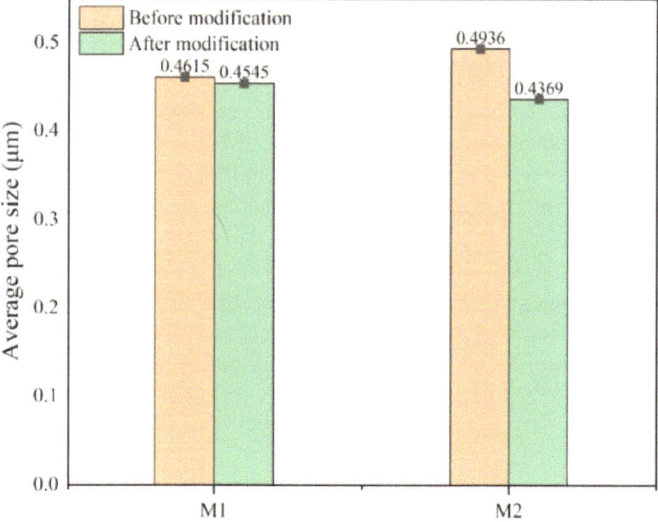

Fig. 5.6 Average pore sizes of M1 and M2 before and after the modification. Reprinted with the permission from Ref. [3] Copyright (2023) (Elsevier)

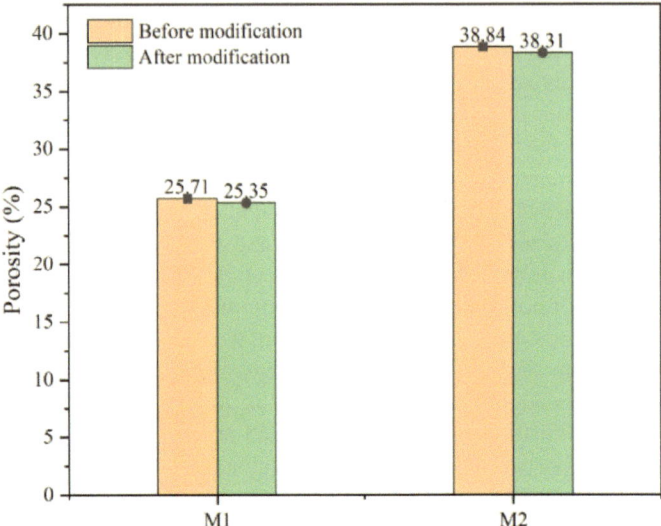

Fig. 5.7 Porosities of M1 and M2 before and after the modification. Reprinted with the permission from Ref. [3] Copyright (2023) (Elsevier)

L/(m^2 min kPa). According to structural parameters of M1 and M2, the membrane thickness of M1 is 1.5 mm, while that of M2 is 2 mm. It can be seen that the membrane thickness has a great influence on the mass transfer performance. In summary, the pore size, porosity and gas flux of the membrane are slightly reduced after the modification, and the modification process has little effect on structures of M1 and M2 membranes.

In order to explore changes in the surface wettability of the ceramic membrane, contact angles of M1 and M2 are measured, as shown in Fig. 5.9. The M1 contact angle increases from 78.9° to 168.0°, while the M2 contact angle increases from 55.8° to 170.7°. After the modification, the wettability of M1 and M2 can be significantly improved, and water droplets cannot penetrate into the pores on inner surfaces of M1 and M2. Therefore, it is possible to ensure that the modified M1 and M2 are not wetted during the CO$_2$ capture experiment with the membrane contactor.

In order to observe the superhydrophobicity of M1 and M2 more directly, the dynamic characteristics of droplets impact on the surface of M1 and M2 are captured by a high-speed camera, as show in Fig. 5.10. Due to the hydrophilicity of original membranes, water droplets begin to spread immediately upon contact with the surface of M1 and M2, and begin to shrink after reaching the maximum spreading

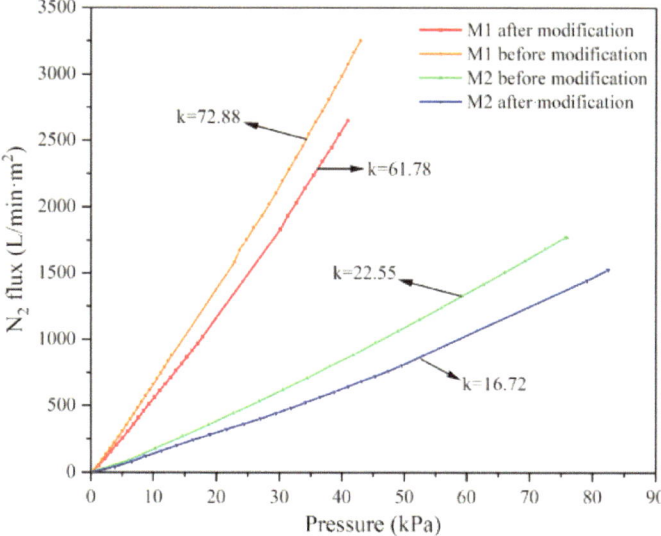

Fig. 5.8 N$_2$ fluxes of M1 and M2 before and after the modification. Reprinted with the permission from Ref. [3] Copyright (2023) (Elsevier)

diameter. In this process, due to the porous nature of the ceramic membrane, water droplets quickly penetrate into membrane holes. For the modified M1 and M2, water droplets begin to spread after touching surfaces of M1 and M2, and begin to shrink after reaching the maximum diameter. Due to the superhydrophobicity of modified membranes, water droplets have less viscous force on the surface. When the droplet retract to a certain extent and bounce from the membrane surface, it rolls off the membrane.

5.5 Experimental Results of CO$_2$ Capture Performance with Ceramic Membranes

5.5.1 Effect of Gas and Absorbent Flowrate on CO$_2$ Capture Performance

In order to avoid the influence of structure parameters of the ceramic membrane contactor on the CO$_2$ capture performance, the *Re* number is used to replace gas and absorber flowrates. Figure 5.11 shows gas mass transfer properties of ceramic membrane M1 and M2 at different *Re* numbers before and after the modification. On both flue gas and absorber sides, *Re* numbers are less than 2300, so flue gas and absorber flows belong to the laminar flow in the membrane contactor. The CO$_2$

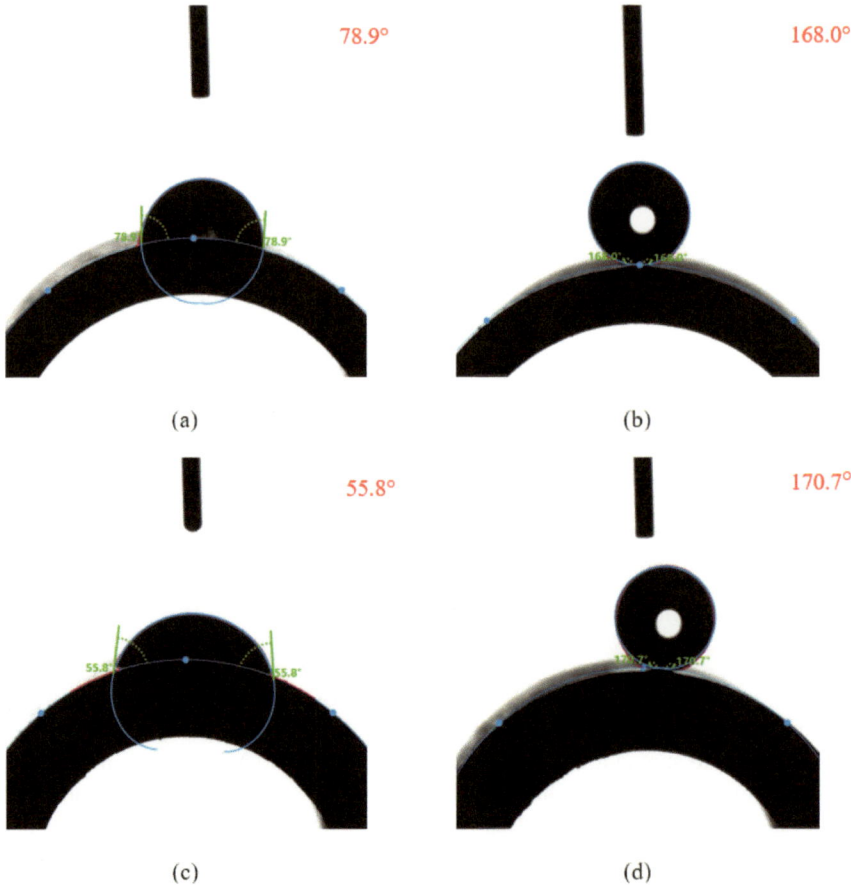

Fig. 5.9 Contact angles of M1 and M2 before and after the modification (**a** M1 before the modification; **b** M1 after the modification; **c** M2 before the modification; **d** M2 after the modification). Reprinted with the permission from Ref. [3] Copyright (2023) (Elsevier)

capture performance of M1 and M2 is significantly improved after the modification. In terms of the mass transfer performance, after the modification, the mass transfer rate of M1 is increased from 5.67×10^{-4} to 14.59×10^{-4} m/s, while the mass transfer rate of M2 is increased from 4.34×10^{-4} to 13.10×10^{-4} m/s. In terms of the capture efficiency, M1 increases from 79.9 to 99.55%, while M2 increases from 78.4 to 98.55%. The mass transfer resistance of the gas is mainly determined by the wetting degree of the membrane. Even if a small part of the membrane is wet, the gas mass transfer flux will drop sharply. The greater the degree of the membrane wetting, the higher the mass transfer resistance. For the modified M1 and M2, liquid breakthrough pressures are 0.63 MPa and 0.65 MPa, respectively. During experiments, the pressure difference between the flue gas side and the absorber side is always lower than 0.1 MPa, M1 and M2 are difficult to be wetted, and the gas will

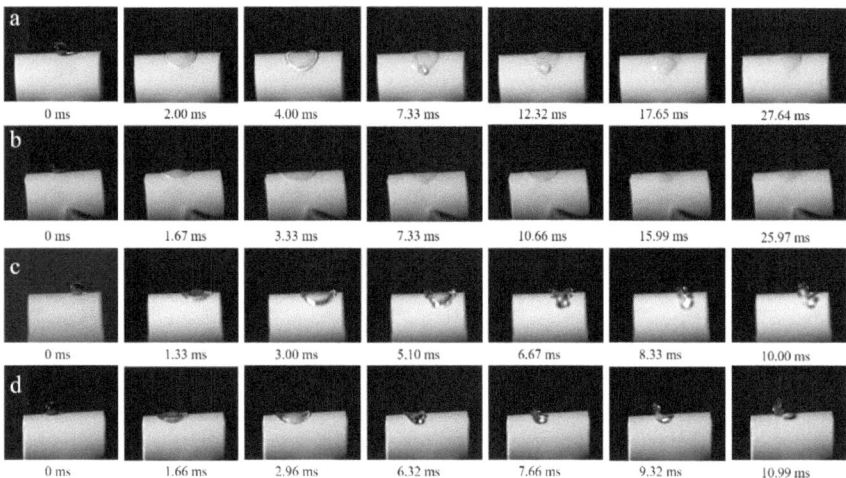

Fig. 5.10 Dynamic characteristics of water droplets impacting surfaces of M1 and M2 before and after the modification (**a** M1 before the modification; **b** M2 before the modification; **c** M1 after the modification; **d** M2 after the modification). Reprinted with the permission from Ref. [3] Copyright (2023) (Elsevier)

fill the entire membrane. Under the action of the concentration difference on both sides of the membrane, the gas diffuses to the absorbent side, and the mass transfer coefficient is mainly determined by the liquid phase mass transfer coefficient. After the modification, mass transfer resistances of M1 and M2 decrease significantly, and capture efficiencies increase.

Figure 5.12 depicts the process of the gas mass transfer within the membrane. Both the flue gas and the absorber are in laminar flow state in the ceramic membrane contactor, and the fluid viscosity hinders the mass transfer process of the flue gas across the membrane. When gas and liquid flowrates increase, the thickness of the boundary layer on both sides of the membrane decreases, and the mass transfer resistance decreases, thus improving the mass transfer performance. In addition, due to the rapid reaction of CO_2 and MEA absorbent in the membrane contactor, the absorber flowrate increases. Moreover, the reaction products at the liquid film interface are quickly taken away, which promotes the absorption reaction, thus improving the mass transfer performance. It is worth noting that although increasing the Re number of the gas phase can also improve the mass transfer coefficient of the membrane, the residence time of the gas in the membrane contactor is shortened due to the increase of the gas flowrate, and the CO_2 capture efficiency reduces.

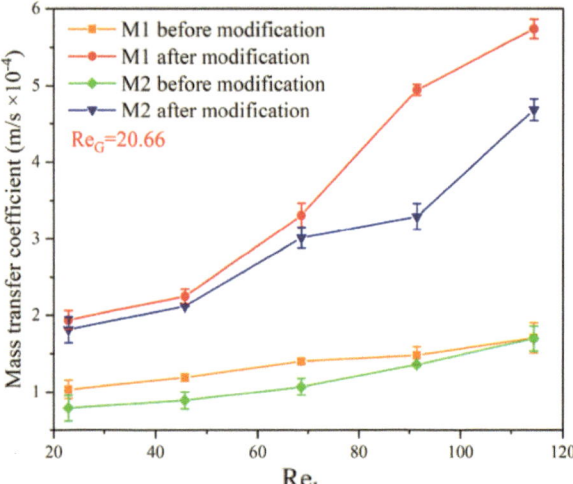

(a) Effect of the *Re* number of the liquid phase o   n the mass transfer coefficient(the *Re*number of the gas phase is 20.66).

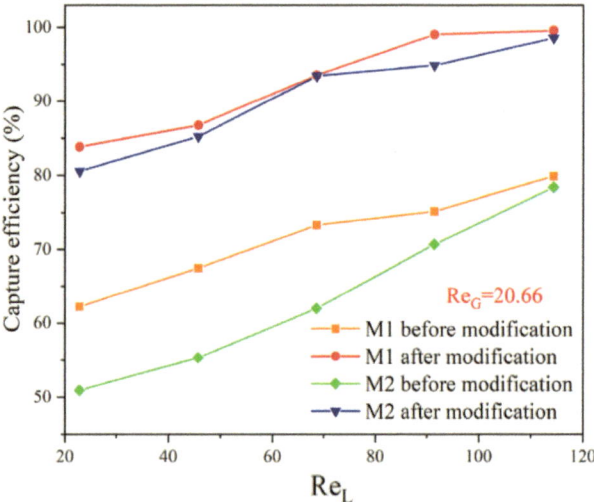

(b) Effect of the *Re*number of the liquid phase on the capture efficiency(the *Re*number of the gas phase is 20.66).

Fig. 5.11 Effect of the *Re* number on the carbon capture performance. Reprinted with the permission from Ref. [3] Copyright (2023) (Elsevier)

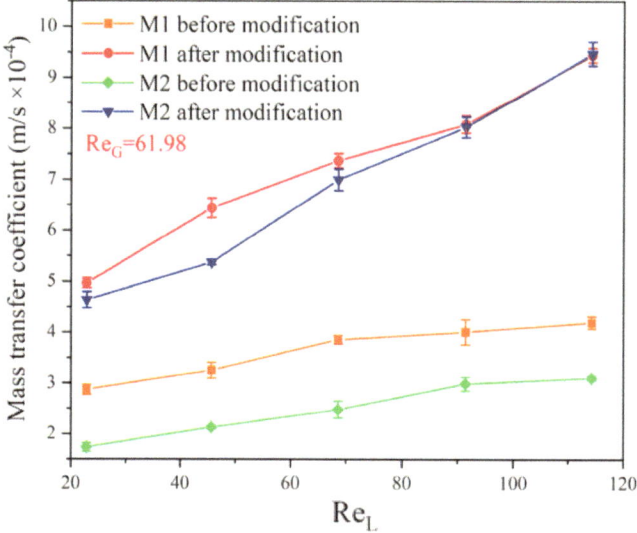

(c) Effect of the *Re* number of the liquid phase on the mass transfer coefficient (the *Re* number of the gas phase is 61.98).

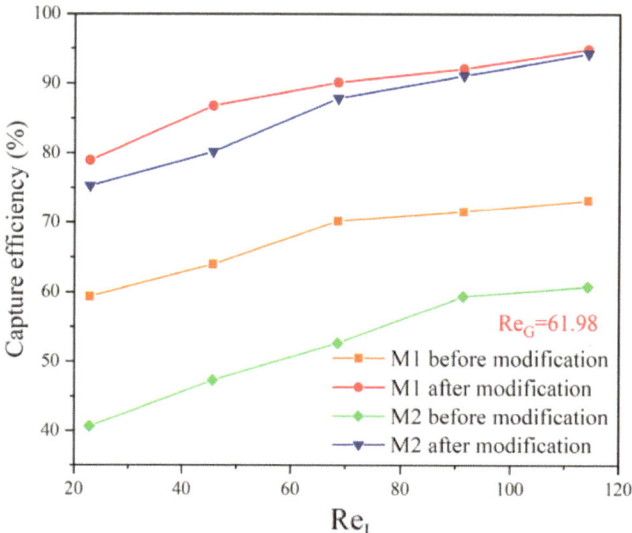

(d) Effect of the *Re* number of the liquid phase on the capture efficiency (the *Re* number of the gas phase is 61.98).

Fig. 5.11 (continued)

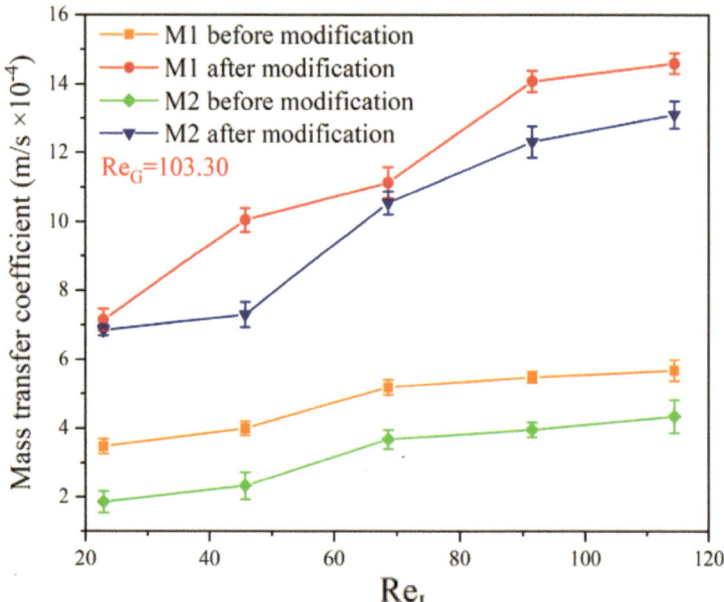

(e) Effect of the   *Re* number of the liquid phase on the mass transfer coefficient (the *Re* number of the gas phase is 103.30).

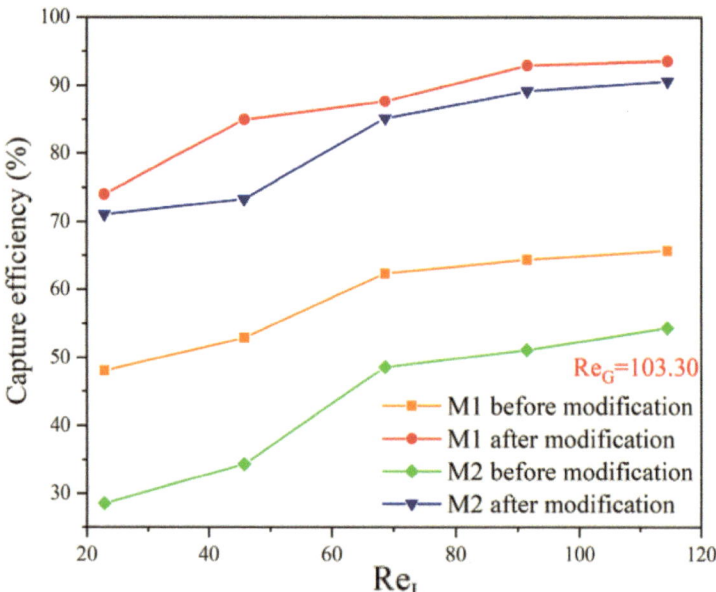

(f) Effect of the *Re* number of the liquid phase on the capture efficiency (the *Re* number of the gas phase is 103.30).

Fig. 5.11 (continued)

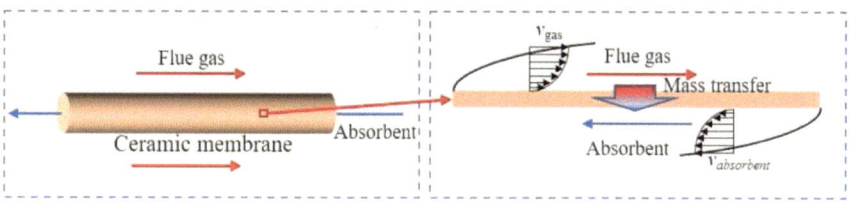

Fig. 5.12 Mass transfer mechanism of the membrane in the laminar flow. Reprinted with the permission from Ref. [3] Copyright (2023) (Elsevier)

5.5.2   Analysis of Mass Transfer Resistance in Ceramic Membranes

Figure 5.13 shows overall mass transfer resistances of M1 and M2 ceramic membrane contactor in the non-wetting state. Under normal circumstances, in the process of CO_2 capture in the membrane, the liquid phase and the membrane phase play leading roles in the mass transfer resistance, while the gas mass transfer resistance can be ignored [9]. At the same absorbent flowrate, overall mass transfer resistances of M1 and M2 ceramic membrane contactors are basically the same. With the increase of *Re* number, mass transfer resistances in the membrane contactor change in the same way and decrease gradually. The overall mass transfer resistance of M1 decreases from 5166.60 to 1743.26 s/m, while the overall mass transfer resistance of M2 decreases from 5523.16 to 2136.05 s/m. In addition, with the increase of the liquid *Re* number, the mass transfer resistance of the gas phase and the membrane phase decrease gradually. M1 decreases from 53.67 to 12.83%, while M2 decreases from 58.78 to 32.82%. It can be seen that the liquid phase resistance in the laminar flow state plays a dominant role in the mass transfer performance of the ceramic membrane during the whole process of CO_2 capture by the membrane contactor. In the actual industrial application process, the absorber flowrate should be increased as much as possible to reduce the mass transfer resistance caused by the liquid boundary layer, thus improving the efficiency of the membrane contactor.

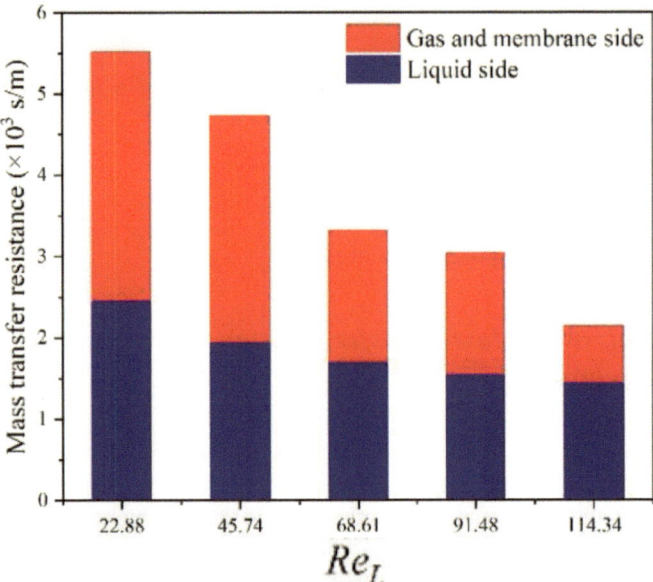

(a) Mass transfer resistance of the modified M1.

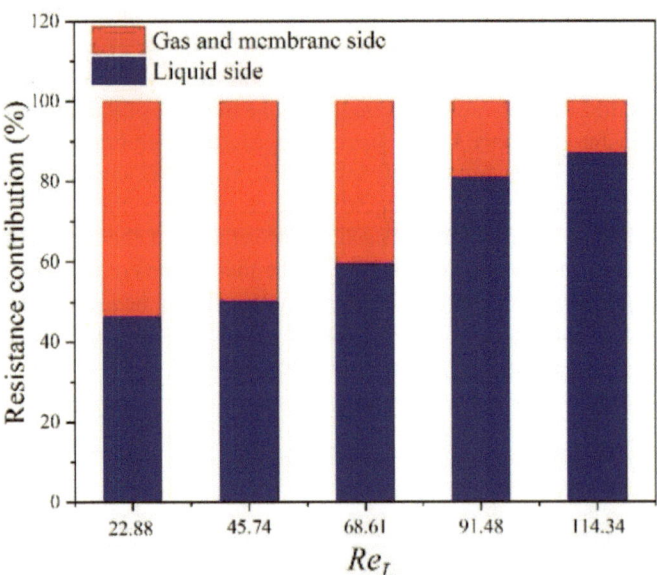

(b) Resistance distribution ratio of the modified M1.

Fig. 5.13 Mass transfer resistance distribution of the modified M1 and M2 under non-wetting states. Reprinted with the permission from Ref. [3] Copyright (2023) (Elsevier)

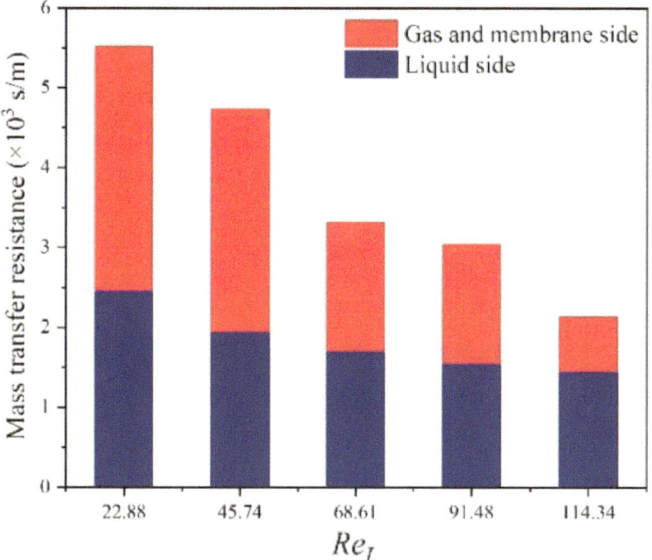

(c) Mass transfer resistance of the modified M2.

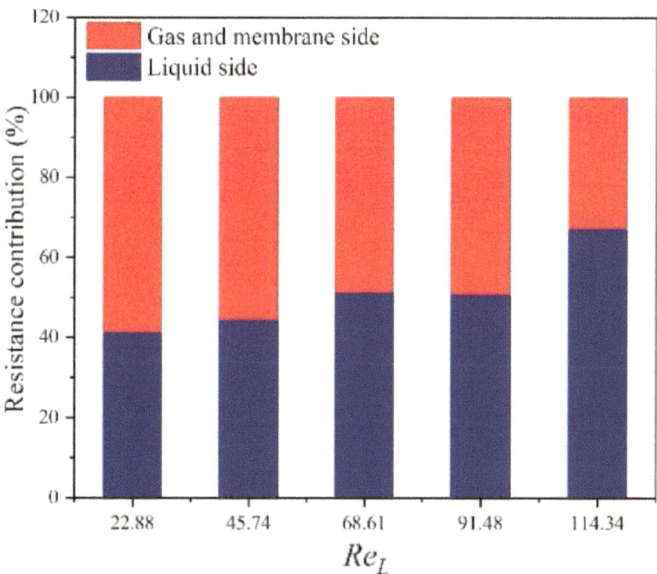

(d) Resistance   distribution ratio of the modified M2.

Fig. 5.13   (continued)

References

1. Lee HJ, Park YG, Kim MK et al (2019) Study on CO_2 absorption performance of lab-scale ceramic hollow fiber membrane contactor by gas/liquid flow direction and module design. Sep Purif Technol 220:189–196
2. Ramli NA, Hashim NA, Aroua MK (2020) Supported ionic liquid membranes (SILMs) as a contactor for selective absorption of CO_2/O_2 by aqueous monoethanolamine (MEA). Sep Purif Technol 230:115849
3. Fu HM, Shen YB, Li ZH et al (2023) CO_2 capture using superhydrophobic ceramic membrane: Preparation and performance analysis. Energy 282:128873
4. Swati IK, Sohaib Q, Khan H et al (2022) Non-dispersive solvent absorption of post-combustion CO_2 in membrane contactors using ionic liquids. J Mol Liq 351:118566
5. Atchariyawut S, Feng C, Wang R et al (2006) Effect of membrane structure on mass-transfer in the membrane gas-liquid contacting process using microporous PVDF hollow fibers. J Membr Sci 285:272–281
6. Zhao SF, Feron PM, Deng LY et al (2016) Status and progress of membrane contactors in post-combustion carbon capture: a state-of-the-art review of new developments. J Membr Sci 511:180–206
7. Cao F, Gao HX, Ling H et al (2020) Theoretical modeling of the mass transfer performance of CO_2 absorption into DEAB solution in hollow fiber membrane contactor. J Membr Sci 593:117439
8. Hassan MA, Mohamed SK, Ibrahim AA et al (2017) A comparative study of the incorporation of TiO_2 into MCM-41 nanostructure via different approaches and its effect on the photocatalytic degradation of methylene blue and CO oxidation. React Kinet Mech Cat 120:791–807
9. Jin PR, Huang C, Shen YD et al (2017) Simultaneous separation of H_2S and CO_2 from biogas by gas liquid membrane contactor using single and mixed absorbents. Energy Fuels 31:11117–11126

Chapter 6
Preparation and Performance Study of Low-Cost Fly Ash Based Superhydrophobic Ceramic Membranes

In this chapter, using the fly ash in the power plant as the main raw material, plate and tubular ceramic membranes are prepared, and then the hydrophobic modification is carried out on them. The flat ceramic membrane is used to investigate the surface characteristics, and the tubular ceramic membrane is used to analyze the CO_2 capture performance. Finally, the economy of the fly ash based ceramic membrane is analyzed, in order to provide technical support for the large-scale application.

6.1 Preparation of Superhydrophobic Fly Ash Ceramic Membranes

In order to analyze the surface characteristics and CO_2 capturing performance of membranes, tubular and flat fly ash based ceramic membranes are prepared by using the same mud material. First, the fly ash, the carboxymethyl cellulose, the dextrin and the glycerin are mixed according to the mass ratio of 89.5: 6: 3: 2.5. Second, adding 400 mL deionized water and mixing well with a mixer. Third, the mixed mud is put into the vacuum mud machine for mud training, and the mud is stale for 24 h after mud training. Fourth, tubular and flat ceramic membrane billets are prepared respectively and dried at room temperature for 48 h. For the tubular ceramic membrane, the mud material is extruded by extruder, and the tubular ceramic membrane body with an outer diameter of 12 mm and an inner diameter of 8 mm is formed initially. Placing it on a straightening machine and drying it for 48 h. For the plate ceramic membrane, 2.5 g mud material is weighed each time and added to the tablet press to make a sheet with a diameter of 20 mm and a thickness of 2 mm, and the tablet pressure is 10 MPa. Fifth, the prepared billet is put into the tube furnace, sintered at a heating rate of 2 °C/min, and held for 3 h after reaching 1150 °C. Finally, tubular and flat ceramic membranes are cooled at room temperature, as shown in Fig. 6.1. The tubular fly ash based ceramic membrane is used to verify the CO_2 capturing performance, and the

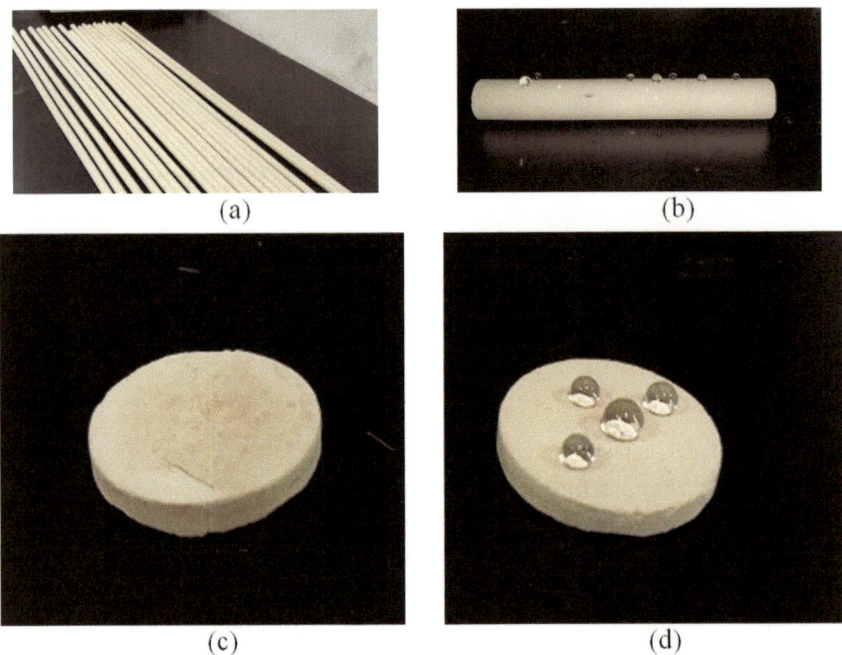

(a)   (b)

(c)   (d)

Fig. 6.1 Tubular and flat fly ash based ceramic membranes (**a** tubular original fly ash based ceramic membrane; **b** tubular fly ash based ceramic membrane after modification with hydrophobicity; **c** flat original fly ash based ceramic membrane; **d** flat fly ash based ceramic membrane after modification with hydrophobicity)

plate fly ash based ceramic membrane is used to measure the surface contact angle and the wettability.

Adopting the dipping method, the low surface energy material is grafted onto the surface of the fly ash based ceramic membrane to achieve the hydrophobic modification of the surface, as shown in Fig. 6.2. First, surface pretreatment. After sintering, the ceramic membrane based on the fly ash is ultrasonic cleaned in the deionized water, the acetone and the anhydrous ethanol for 10 min, respectively. Then taking it out and drying it in an oven at 80 °C for 6 h to fully dry the ceramic membrane. After drying, cooling it to room temperature for the use. Second, preparing the modified solution. An appropriate amount of anhydrous ethanol and 1H,1H,2H,2H-perfluorodecyl triethoxysilane are utilized for preparing the modified solution. After preparing, the solution should be left for 12 h at the room temperature. Third, surface grafting modification. The pre-treated fly ash based ceramic membrane is immersed in the above modified solution. Then removing and fully cleaning with the deionized water and fully drying it in an oven at 80 °C to obtain the superhydrophobic fly ash based ceramic membrane.

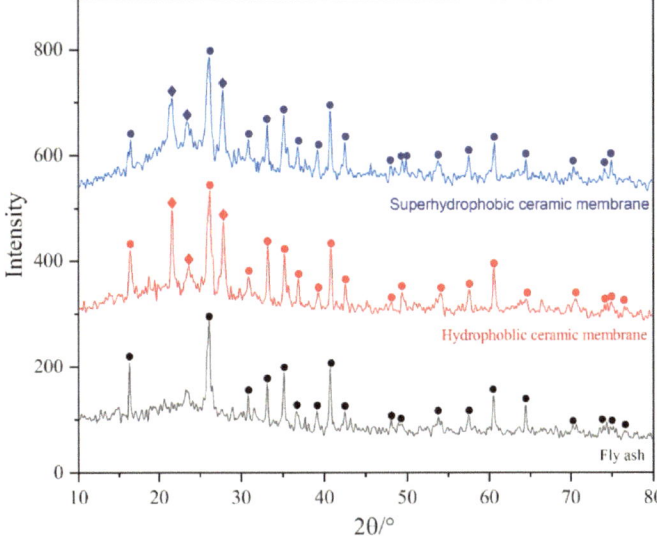

Fig. 6.2 Phase analysis of the fly ash, fly-ash-based ceramic membranes before and after the modification. Reprinted with the permission from Ref. [1] Copyright (2022) (Elsevier)

6.2   Performance Analysis of Fly Ash Based Superhydrophobic Membranes

XRF is used to measure the chemical composition of the fly ash, and results are shown in Table 6.1. It can be seen that main components of the fly ash are SiO_2 and Al_2O_3. In addition, it also contains a small amount of Fe_2O_3, CaO, MgO, TiO_2 and so on.

Table 6.1 Chemical composition of the fly ash

Composition	Content (wt%)
SiO_2	35.3
Al_2O_3	35.8
Fe_2O_3	3.6
CaO	4.38
MgO	0.55
TiO_2	1.91
CO_2	15
SO_3	1.08
Others	2.38

In order to determine the crystal phase changes of materials in each step of the preparation process, the fly ash, the ceramic base membrane and the superhydrophobic ceramic membrane are analyzed by the XRD diffraction (Fig. 6.2). The main crystalline phase of the fly ash is the mullite phase (dots in Fig. 6.2), and the diffraction peak at its small angle has a slight bulge, indicating that the fly ash also contains a lot of amorphous glass phases. The phase composition of ceramic base membrane and the superhydrophobic ceramic membrane cannot change greatly. Its main crystalline phase is still the mullite phase, in addition to a small amount of the calcium feldspar phase (rhomboids in Fig. 6.2). After sintering, the diffraction peak of the crystalline phase of the membrane at a small angle is similar to that of the fly ash, and also shows a slight uplift state, indicating that the membrane also contains a large number of amorphous glass phases.

Figure 6.3 shows surface morphologies of the ceramic base membrane and the superhydrophobic ceramic membrane. Fly ashes on the surface of the ceramic membrane before and after the modification are sintered at a high temperature and bonded together to form a rough surface structure. Surface morphologies of ceramic membranes do not change significantly before and after the modification. It is mainly due to the fact that fluorosilane molecules only form a thin molecular layer on the surface of the fly ash based ceramic membrane, which only reduces the surface free energy and does not change the surface structure of the membrane.

In order to ensure that 1H,1H,2H,2H-perfluorodecyl triethoxysilane is grafted onto the fly ash substrate surface, the energy dispersive spectroscopy is performed on ceramic membrane surfaces (Fig. 6.4). After the modification, elements on the membrane surface are basically the same, and the absorption peak of the fluorine appears on the membrane surface. The element is only present in the grafted substance 1H,1H,2H,2H-perfluorodecyl triethoxysilane, so it can be determined that 1H,1H,2H,2H-perfluorodecyl triethoxysilane has been successfully grafted to the surface of the fly ash based ceramic membrane.

In order to quantitatively analyze the influence of surface wetting characteristics on the dynamic behavior of droplet impact, contact angles of ceramic membranes

(a)   (b)

Fig. 6.3 Surface morphologies of fly ash based ceramic membranes (**a** before the modification; **b** after the modification) Reprinted with the permission from Ref. [1] Copyright (2022) (Elsevier)

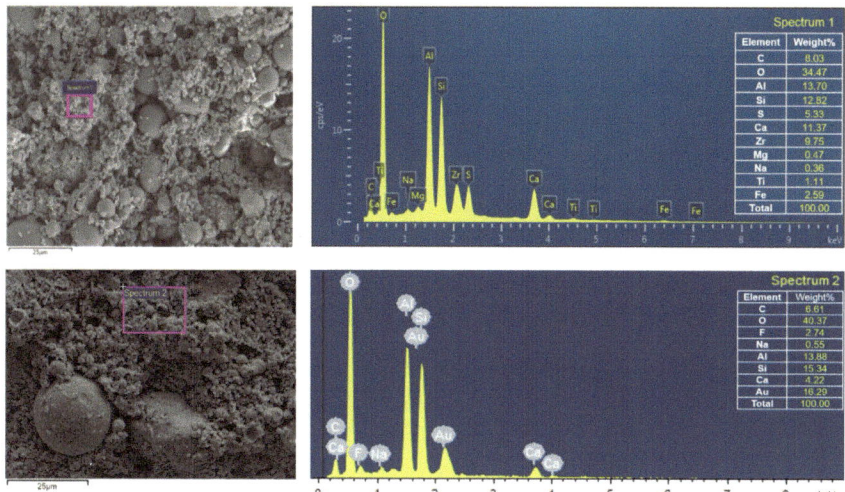

Fig. 6.4 Energy dispersive spectroscopy of fly ash-based ceramic membrane surfaces (**a** before the modification; **b** after the modification). Reprinted with the permission from Ref. [1] Copyright (2022) (Elsevier)

before and after the modification are measured, shown in Figs. 6.5 and 6.6. The ceramic base membrane shows the hydrophilicity. Due to the existence of the hydroxyl group on the surface and the loose porous structure in the interior, water droplets quickly penetrate into the ceramic membrane after touching the membrane surface. The contact angle decreases rapidly from the initial 45.6°–0° within 0.4 s. Because the graft of fluorosilane molecules onto the surface reduces its surface energy, the ceramic membrane exhibits the superhydrophobicity. Its contact angle is always maintained at about 151°. In experiments, by slowly rotating the angle of the stage, the tilt angle is recorded at the moment when the droplet starts to roll, and the rolling angle of the surface is determined. The rolling angle of the hydrophobic ceramic membrane is determined to be 11.3° by taking the average value of several measurements.

6.3 CO$_2$ Capture Performance of Modified Fly Ash Based Ceramic Membranes

6.3.1 Effect of the Absorber Flowrate

Figure 6.7 shows the effect of the absorbent flowrate on the CO$_2$ capture performance of the superhydrophobic fly ash based ceramic membrane and the superhydrophobic Al$_2$O$_3$ ceramic membrane. CO$_2$ mass transfer coefficients and capture efficiencies of

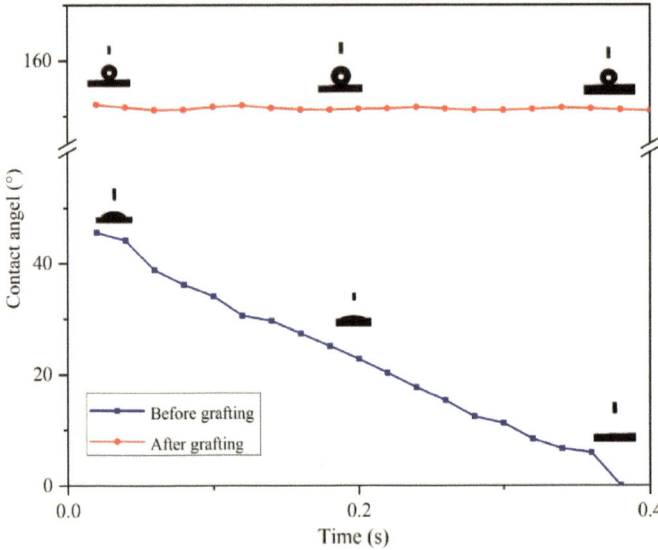

Fig. 6.5 Change trend of contact angles of fly ash based ceramic membranes before and after the modification with time. Reprinted with the permission from Ref. [1] Copyright (2022) (Elsevier)

both membranes increase with the increase of absorbent flowrates. When the flue gas flowrate is 0.6 L/min, the absorbent flowrate increases from 0.02 to 0.1 L/min, the mass transfer coefficient of the superhydrophobic fly ash based ceramic membrane increases from 4.85×10^{-4} to 9.58×10^{-4} m/s, and the CO_2 capture efficiency increases from 76.1 to 95.6%. Correspondingly, the mass transfer coefficient and the CO_2 capture efficiency of the superhydrophobic Al_2O_3 ceramic membrane increase from 4.63×10^{-4} m/s and 75.3% to 9.47×10^{-4} m/s and 94.3%, respectively. Since the flow state of the absorbent in the ceramic membrane is the laminar flow state, the mass transfer process is mainly dominated by the flow boundary layer in the ceramic membrane. When the flowrate of the absorber increases, the thickness of the boundary layer decreases, and the mass transfer resistance between the flue gas and the absorber on both sides of the ceramic membrane decreases. Therefore, the CO_2 mass transfer coefficient in the ceramic membrane increases with the increase of the absorbent flowrate, and the capture efficiency also increases.

6.3.2   Effect of the Gas Flowrate

Figure 6.8 shows the effect of the flue gas flowrate on the CO_2 capturing performance of the superhydrophobic fly ash based ceramic membrane and the superhydrophobic Al_2O_3 ceramic membrane. With the increase of the flue gas flowrate, mass transfer coefficients of both membranes increase, and CO_2 capture efficiencies decrease. When the absorbent flowrate is 0.06 L/min and the flue gas flowrate

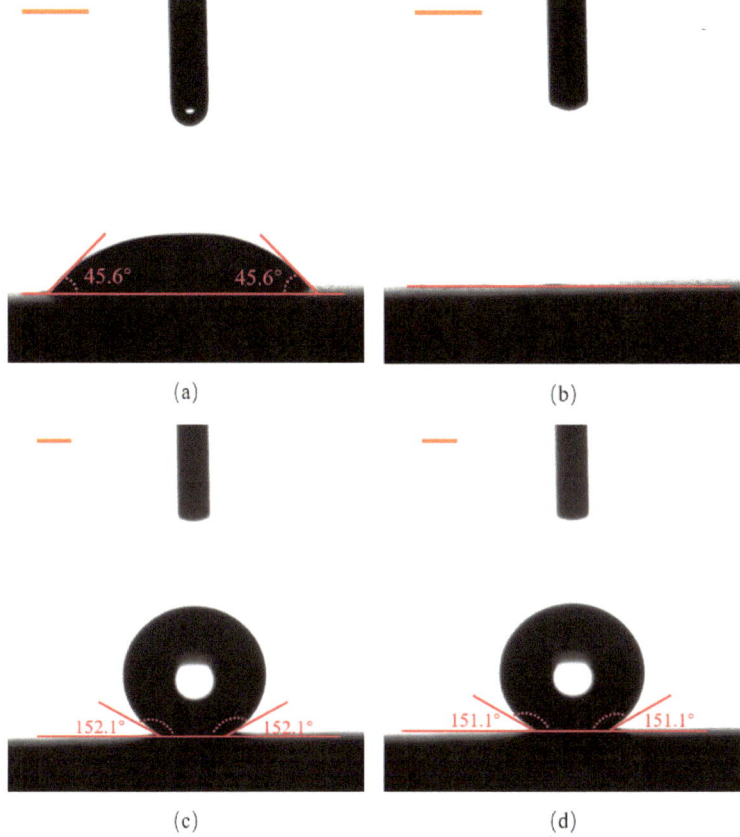

Fig. 6.6 Contact angles of ceramic membranes (**a** before the modification, $t = 0$ s; **b** before the modification, $t = 0.34$ s; **c** after the modification, $t = 0$ s; **d** after the modification, $t = 0.82$ s). Reprinted with the permission from Ref. [1] Copyright (2022) (Elsevier)

increases from 0.2 to 1.0 L/min, the mass transfer coefficient of the fly ash based ceramic membrane increases from 3.22×10^{-4} to 11.86×10^{-4} m/s, and the CO_2 capture efficiency decreases from 94.2 to 86.5%. Correspondingly, the mass transfer coefficient of the superhydrophobic Al_2O_3 ceramic membrane increases from 3.01×10^{-4} to 10.57×10^{-4} m/s, and the capture efficiency decreases from 93.5 to 85.2%. The flow state of the gas in the shell side is also laminar flow. When the flue gas flowrate increases, the thickness of the boundary layer at the boundary layer decreases, the mass transfer resistance decreases, and the mass transfer coefficient increases. However, the increase of the flue gas flowrate leads to the shortening of the contact time between the flue gas and the ceramic membrane. Some flue gas is taken out of the membrane contactor before it has time to transfer into the membrane. Therefore, the CO_2 capture efficiency decreases with the increase of the flue gas flowrate.

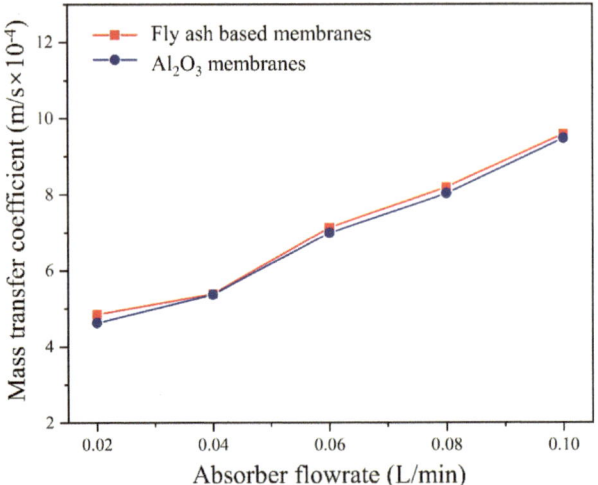

(a) Effect of the absorber flowrate on the mass transfer coefficient.

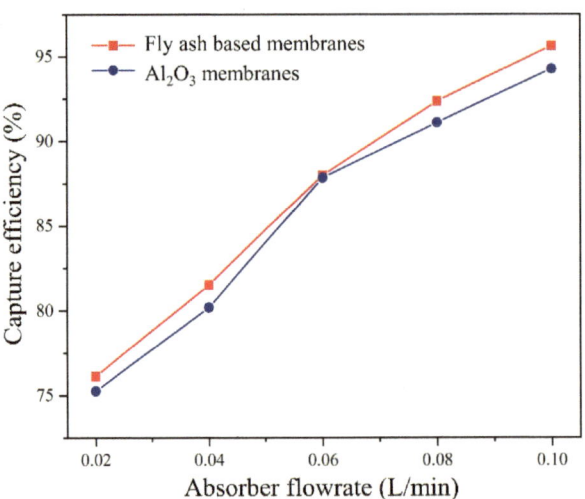

(b) Effect of the absorber flowrate on the capture efficiency.

Fig. 6.7 Effect of the absorber flowrate on the CO_2 capture performance

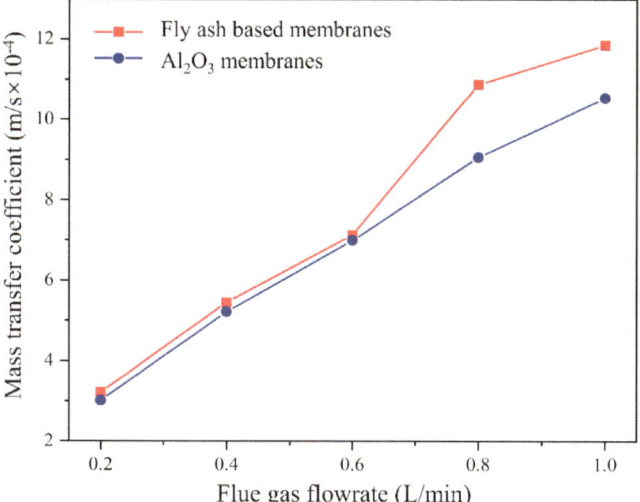

(a) Effect of the flue gas flowrate on the mass transfer coefficient.

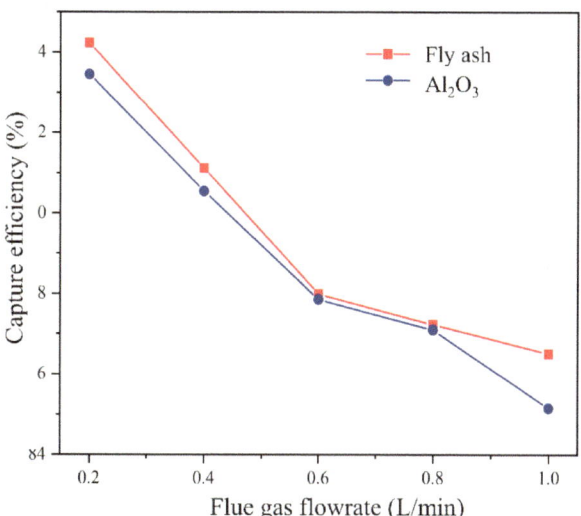

(b) Effect of the flue gas flowrate on the capture efficiency.

Fig. 6.8 Effect of the flue gas flowrate on the CO$_2$ capture performance

6.4 Dynamic Properties of Droplet Impact on Membrane Surfaces

6.4.1 Dynamic Characterization of Droplet Impacts on Membrane Surfaces

The impact process of water droplets on the surface of the fly ash based ceramic membrane is shown in Fig. 6.9. As shown in Fig. 6.9a, due to the hydrophilicity of the surface, the droplet first stacks in a multi-layer cake shape on the surface ($t = 3.507$ ms), and then spreads out to the maximum diameter ($t = 6.513$ ms). Finally, it rebounds slightly ($t = 9.853$ ms). However, the strong viscous force between the droplet and the surface inhibits the rebound of the droplet and sticks to the droplet on the surface. When droplets impact the modified hydrophobic fly ash based ceramic membrane surface, its dynamic characteristics are different from those of the hydrophilic surface (Fig. 6.9b–g). The dynamic behavior of droplets impacting the superhydrophobic surface of the ceramic membrane can be divided into the spreading, the retracting and the partial rebound. When the droplet impacting velocity is low (Fig. 6.9b–d), droplets evenly stack in all directions due to the action of the inertia force. The spreading outline is circular, and the spreading diameter reaches the maximum when it is about 3 ms. After that, the droplet enters the retraction stage and partially rebounds on the hydrophobic surface. The rebound droplet shows a thin shape under the action of the inertia force and the viscous force. With the increase of the impacting velocity (Fig. 6.9e–g), droplets spread to the maximum diameter on the membrane surface and reassemble, resulting in the splash. The kinetic energy obtained by some droplets is greater than the surface energy, causing some droplets to continue to spread around in the shape of petals, and the rest of droplets to shrink to the center. In the process of the droplet retraction, the interaction force among droplets is less than that between the droplet and the surface. As a result, only a part of droplets reassemble and rebound on the surface. The rest of droplets splashes away from the center, forming numerous small discrete droplets on the surface.

6.4.2 Solid–Liquid Contact Time

In the study of dynamic characteristics of droplets impact on the superhydrophobic surface, the solid–liquid contact time directly affects the energy exchange process between droplets and the surface. Figure 6.10 shows the spread time, the retraction time, and the total contact time under different impact velocities. When other conditions are fixed, the above time does not change with the impact velocity. The average spreading time, the average retraction time and the average total contact time are 2.9, 11.5 and 14.4 ms, respectively. In addition, the droplet retraction time is much longer than the spreading time.

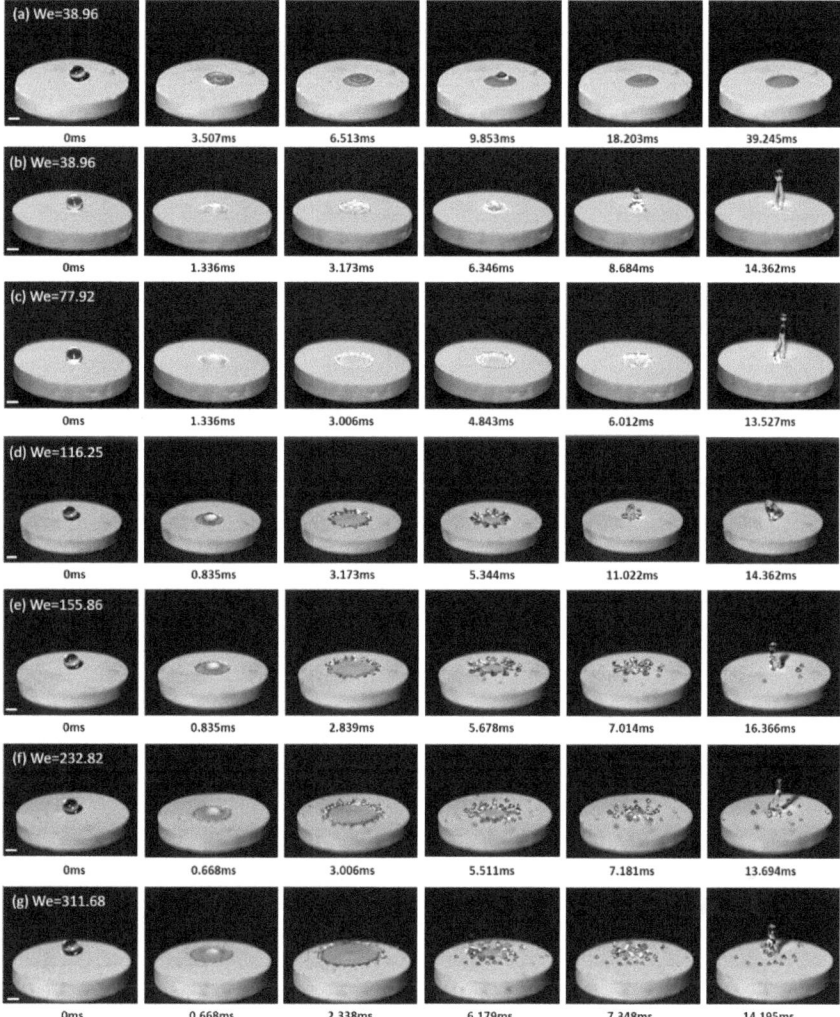

Fig. 6.9 Dynamic characteristics of droplets impacting on the membrane surface (**a** hydrophilic surface; **b–g** superhydrophobic surface; scale bars are 2 mm). Reprinted with the permission from Ref. [1] Copyright (2022) (Elsevier)

6.4.3 Contact Diameter

Figure 6.11 shows the effect of impact velocities on the contact diameter. After the droplet hits the hydrophobic surface, the droplet rapidly expands to its maximum diameter at about 2.9 ms, and then rebounces. The maximum spread diameter increases with the increase of the collision velocity. In the collision process, the time of the droplet spreading stage is much smaller than the retraction time, and the

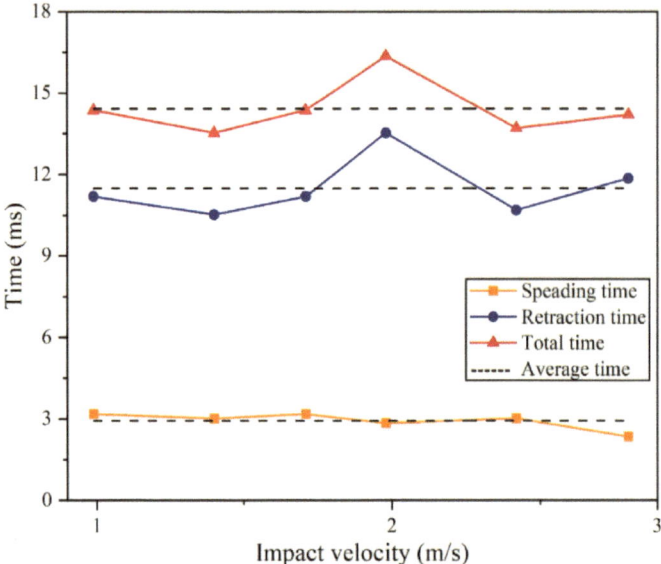

Fig. 6.10 Relationship between times of droplet impacting velocities. Reprinted with the permission from Ref. [1] Copyright (2022) (Elsevier)

spreading rate is much larger than the retraction rate. When the droplet spreads to the maximum diameter and begins to shrink, the shrinkage process can be divided into the fast shrinkage process and the slow shrinkage process, and curves of fast shrinkage and spreading processes are almost mirror symmetric [2].

6.4.4 Maximum Spreading Factor

The variation of the maximum spread coefficient ($\beta_{\max}$) with the *We* number is often used to study the dynamic behavior of droplets impact the superhydrophobic surface [3]. Figure 6.12 shows the change of $\beta_{\max}$ with $We^{1/4}$. Through fitting, it can be seen that $\beta_{\max}$ increases linearly with the increase of $We^{1/4}$, and the fitted linear equation is as follows.

$$\beta_{\max} = 0.92We^{1/4} \tag{6.1}$$

The correlation coefficient of Eq. (6.1) is 0.9952, which means that the droplet spreading process on the superhydrophobic surface is mainly determined by the inertia force and the capillary force of the droplet hitting the superhydrophobic surface. In previous studies, Wang et al. [4] studied the dynamic spreading characteristics of droplets impact on the superheated superhydrophobic surface, and found that the maximum spreading coefficient increases with the increase of the *We* number.

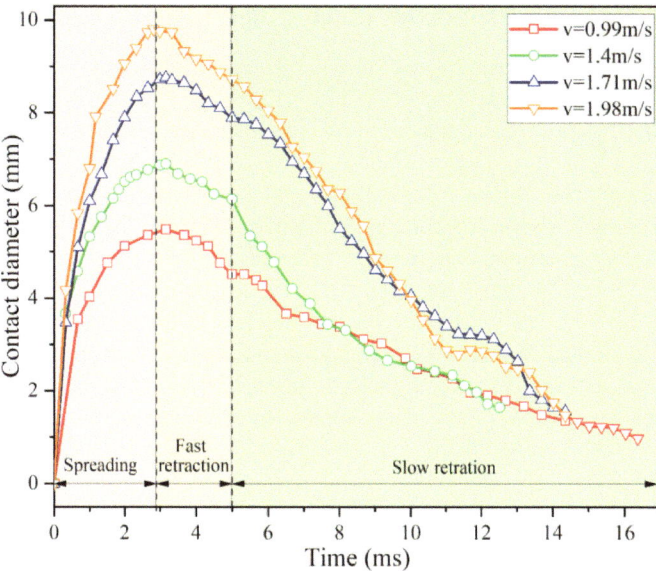

Fig. 6.11 Relationship between times and contact diameters. Reprinted with the permission from Ref. [1] Copyright (2022) (Elsevier)

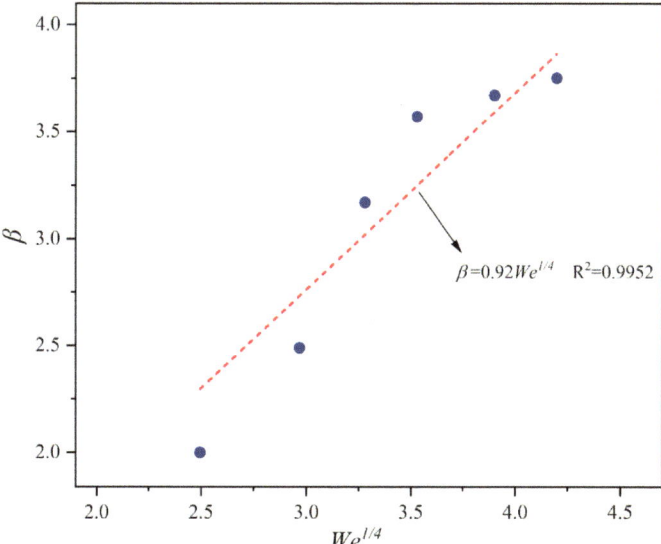

Fig. 6.12 Relationship between β and $We^{1/4}$. Reprinted with the permission from Ref. [1] Copyright (2022) (Elsevier)

6.5   Self-cleaning Properties of Modified Fly Ash Ceramic Membranes

Considering the practical application of the modified fly ash based ceramic membrane, its self-cleaning performance is qualitatively evaluated in this section. The experiment process is as follows. First, 200 g/L fly ash suspension is prepared (Fig. 6.13). Then ceramic membranes are immersed in the fly ash suspension for 10 min. Finally, ceramic membranes are removed and the surface contamination is observed. Figure 6.14 shows the results. Due to the hydrophilicity of the surface, a large number of fine fly ash particles are attached to the surface of the ceramic membrane, which causes the serious pollution. Even after cleaning and drying, the surface cannot recover. When the superhydrophobic ceramic membrane is immersed in the fly ash suspension, the fly ash cannot be fixed on the surface. After removal, there is no visible contamination on the surface, showing excellent self-cleaning performance.

In order to further study the self-cleaning performance of the superhydrophobic fly ash based ceramic membrane, a high-speed camera method is used to observe the dynamic behavior characteristics of droplets containing the fly ash on the membrane surface, and the results are shown in Fig. 6.15. The whole dynamic process is divided into spreading, retracting and partial rebound stages. Under experimental conditions, the average spreading time to the maximum diameter of fly ash droplets after impacting the surface is 3.2 ms, which is 0.3 ms longer than the average contact time

Fig. 6.13 Fly ash suspension with concentration of 200 g/L

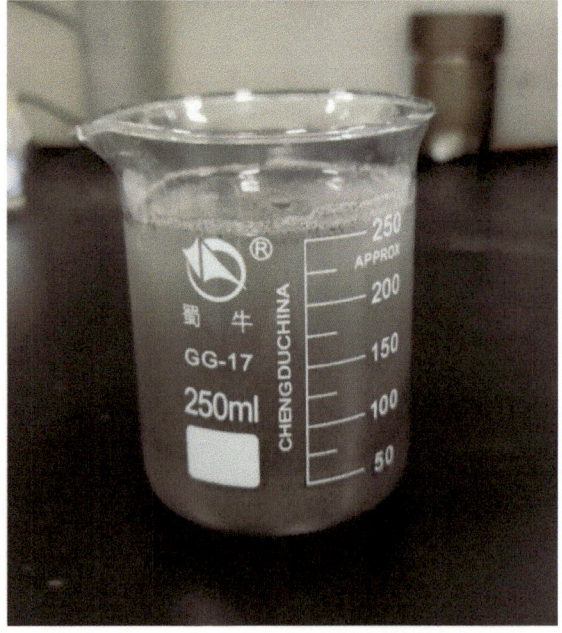

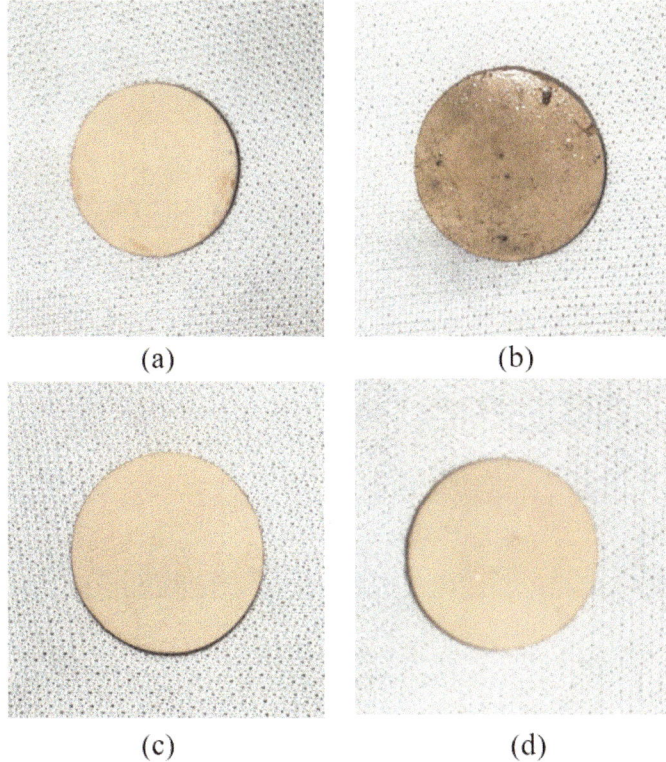

(a)   (b)

(c)   (d)

Fig. 6.14 Comparison of the self-cleaning performance between fly ash based ceramic membranes before and after the modification (**a** membrane before the modification; **b** membrane before the modification soaked in the fly ash suspension; **c** superhydrophobic membrane after the modification; **d** superhydrophobic membrane after the modification soaked in the fly ash suspension)

of the deionized water impacting the surface. The average solid–liquid contact time is 16.032 ms, which is 1.632 ms higher than that of the deionized water impacting the surface. The maximum spread diameter of the droplet impacting surface increases with the increase of the impact velocity. In summary, when the droplet contains impurities, the contact time between the droplet and the superhydrophobic surface increases. Fly ash droplets are more likely to adhere to the superhydrophobic surface, without taking advantage of the self-cleaning property of the superhydrophobic ceramic membrane.

Figure 6.16 shows the maximum spread diameter of the droplet. When the We number is lower than 200, the maximum spreading diameter of the deionized water impacting the surface is always higher than that of droplets containing the fly ash. When the We number is higher than 200, the result is reversed. When the fly ash is added to the deionized water, the droplet viscosity increases. At the low We number, the viscous force predominates during the droplet collision. The increase of the droplet viscosity will inhibit the droplet spreading, and the maximum spreading

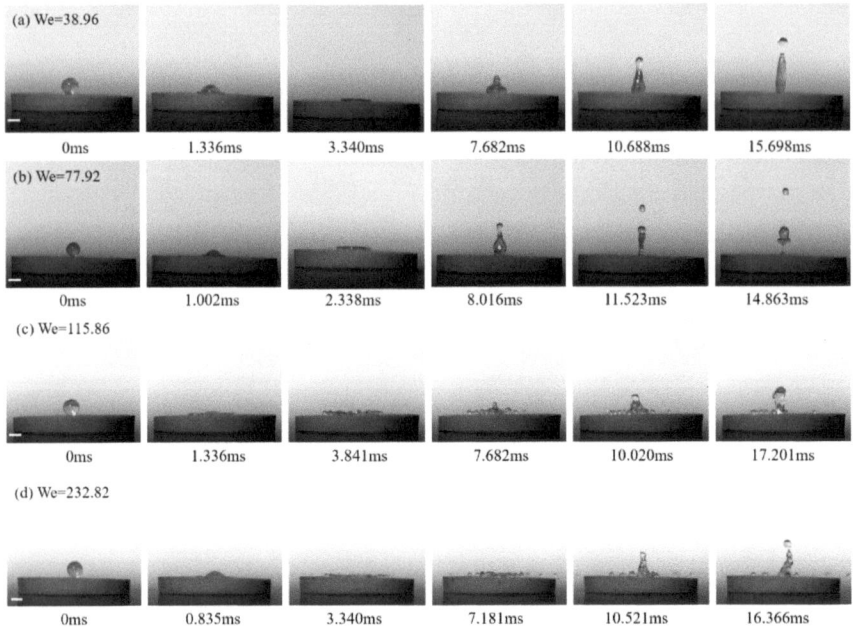

Fig. 6.15 Dynamic characteristics of droplets containing the fly ash impacting the membrane surface (scale bar is 2 mm). Reprinted with the permission from Ref. [1] Copyright (2022) (Elsevier)

diameter of the droplet containing fly ash impacting the surface is smaller than that of the deionized droplet. At the high *We* number, the inertial force dominates the droplet collision process, and the viscosity has little influence on the droplet spreading. The maximum spreading diameter of the droplet containing the fly ash impacting the surface is larger than that of the deionized droplet.

In order to verify the corrosion resistance of the surface of the modified super-hydrophobic ceramic membrane, H_2SO_4 solution with pH = 2 and NaOH solution with pH = 13 are dropped onto the membrane surface respectively, and the contact angle changes are shown in Figs. 6.17 and 6.18. Contact angles of acidic droplets and alkaline droplets on the superhydrophobic ceramic membrane decrease slightly with the time. The acidic droplet contact angle decreases from the initial 151.0°–150.93° within 5 s. While the contact angle of the alkaline droplet within 5 s decreases from the initial 150.96°–150.57°. Under acidic conditions, the average contact angle of the superhydrophobic ceramic membrane surface is 150.98°. Under alkaline conditions, the average contact angle is 150.78°.

In order to further verify the acid-alkaline resistance of superhydrophobic fly ash based ceramic membranes, they are immersed in pH = 2 H_2SO_4 solution and pH = 13 NaOH solution respectively, and remove at intervals. Contact angles of droplets on the surface are measured after cleaning and drying with the deionized water. After 24 h soaking, contact angles of ceramic membranes decrease to different degrees. After soaking in the H_2SO_4 solution for 24 h, the contact angle decreases from

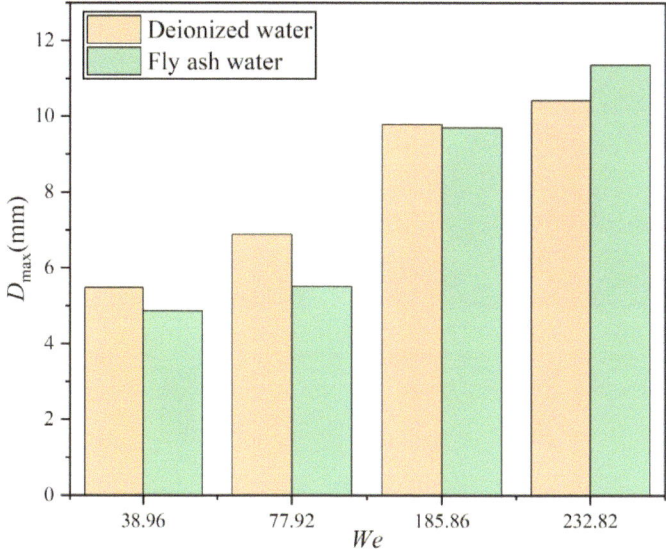

Fig. 6.16 Maximum spreading diameter of the deionized water and the fly ash droplet. Reprinted with the permission from Ref. [1] Copyright (2022) (Elsevier)

150.98° to 148.9°. After soaking in the NaOH solution for 24 h, the contact angle decreases from 150.98° to 146.52°. It shows that superhydrophobic fly ash based ceramic membrane has good acid and alkali resistance.

6.6 Cost Analysis of Fly Ash and Commercial Al₂O₃ Ceramic Membranes

The self-made superhydrophobic fly ash based ceramic membrane shows excellent characteristics in terms of the CO_2 capture performance, the stability and the corrosion resistance. In order to verify the feasibility of the practical application, it is necessary to discuss the economy. So far, no ceramic membrane has been used in engineering applications for CO_2 capture. The relevant study [5] has shown that in the engineering application of ceramic membrane recovering water and waste heat in the flue gas, costs of ceramic membranes accounts for more than 60% of the total investment cost. Therefore, costs of ceramic membranes have a great impact on the large-scale application of CO_2 capture using ceramic membranes. In this section, preparation costs of the Al_2O_3 ceramic membrane and the fly ash based ceramic membrane are analyzed economically, and advantages of fly ash based ceramic membranes in the industrial application are discussed.

Raw material costs per unit membrane area for the preparation of the fly ash based ceramic membrane and the Al_2O_3 ceramic membrane are 5.02 yuan/m² and 27.41

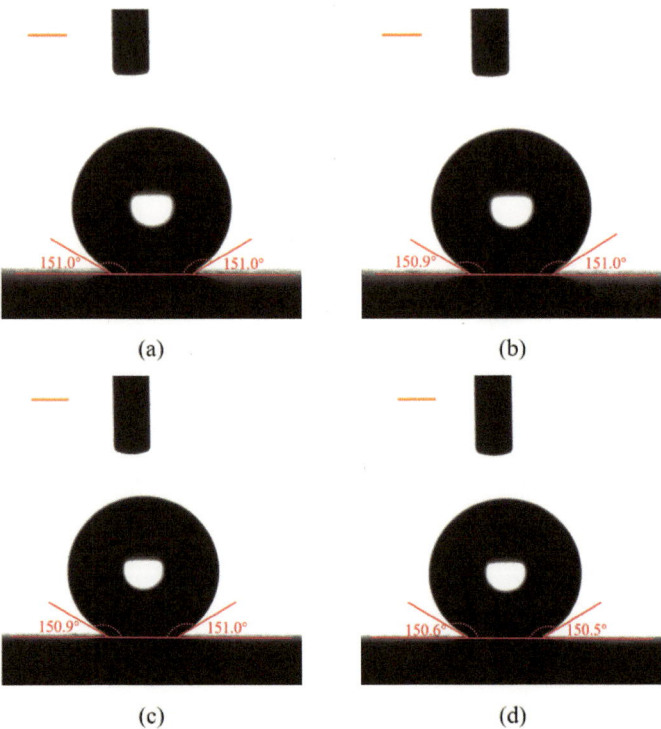

Fig. 6.17 Contact angle of H_2SO_4 solution and NaOH solution on the fly ash based ceramic membrane surface (**a** H_2SO_4 droplet, $t = 0$ s; **b** H_2SO_4 droplet, $t = 4.2$ s; **c** NaOH droplet, $t = 0$ s; **d** NaOH droplet, $t = 4.2$ s; scale bar is 1 mm). Reprinted with the permission from Ref. [1] Copyright (2022) (Elsevier)

yuan/m², respectively. In addition, because the forming of the ceramic membrane requires the high temperature sintering, resulting in the high energy consumption, the sintering temperature of the Al_2O_3 ceramic membrane is usually about 1600 °C. The sintering temperature of the fly ash based ceramic membrane is 1100–1200 °C. In contrast, the sintering energy consumption of the ceramic membrane based on fly ash is lower. Table 6.2 analyzes preparation costs of superhydrophobic fly ash based ceramic membranes and Al_2O_3 ceramic membranes. The cost difference mainly lies in preparation costs of both base membranes, and the rest of preparation costs is basically the same. The cost of the fly ash based superhydrophobic ceramic membrane is 1.68% lower than that of the superhydrophobic Al_2O_3 ceramic membrane. In raw material costs of preparing superhydrophobic ceramic membranes, the cost of the fluorosilane is the highest. Therefore, in the actual application process, under the premise of ensuring the quality of super hydrophobic ceramic membranes, the amount of the fluorosilane should be reduced as much as possible, so as to reduce the overall cost.

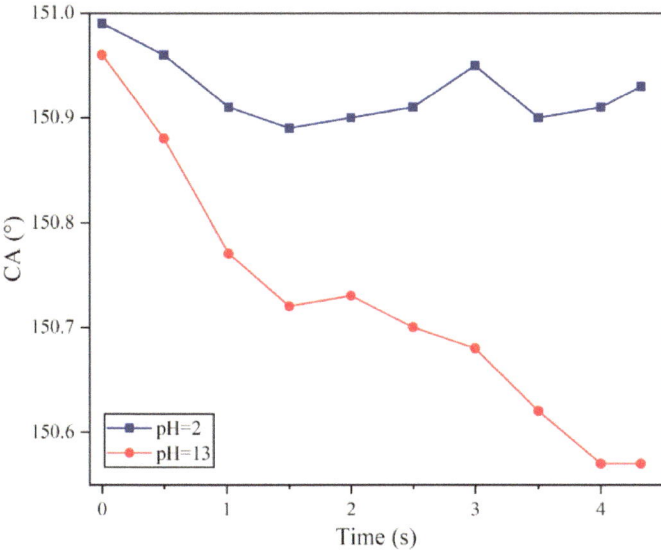

Fig. 6.18 Changes of contact angles of H_2SO_4 solution and NaOH solution on the ceramic membrane with the time. Reprinted with the permission from Ref. [1] Copyright (2022) (Elsevier)

Table 6.2 Preparation cost analysis of the superhydrophobic fly ash based membrane and superhydrophobic Al_2O_3 ceramic membrane

Items	Fly ash based membrane	Al_2O_3 membrane
Cost per unit base-membrane area (Yuan/m²)	5.02	20.74
Fluorosilane unit price (Yuan/g)	10.92	10.92
Amount of fluorosilane per unit membrane area (g/m²)	76.3	76.3
Unit price of anhydrous ethanol (Yuan/g)	0.06	0.06
Amount of anhydrous ethanol per unit membrane area (g/m²)	790	790
Acetone unit price (Yuan/g)	0.09	0.09
Acetone per unit membrane area (g/m²)	395	395
Total cost (Yuan/m²)	919.59	935.31

After the above analysis, fly ash based ceramic membranes have higher advantages than Al_2O_3 ceramic membranes in the CO_2 capture performance and the preparation cost. In addition, developing high-value fly ash products and realizing efficient utilization of the fly ash are conducive to the development process of the comprehensive utilization of the solid waste. Therefore, the application prospect of fly ash based ceramic membranes in the CO_2 capture is excellent, and it is expected to be applied in the large-scale CO_2 capture industry.

References

1. Fu HM, Li ZH, Zhang YT et al (2022) Preparation, characterization and properties study of a superhydrophobic ceramic membrane based on fly ash. Ceram Int 48:11573–11587
2. Li H, Zhang K (2019) Dynamic behavior of water droplets impacting on the superhydrophobic surface: both experimental study and molecular dynamics simulation study. Appl Surf Sci 498:143793
3. Liu H, Si C, Cai C et al (2021) Experimental investigation on impact and spreading dynamics of a single ethanol-water droplet on a heated surface. Chem Eng Sci 229:116106
4. Wang ZF, Xiong JB, Yao WY et al (2019) Experimental investigation on the Leidenfrost phenomenon of droplet impact on heated silicon carbide surfaces. Int J Heat Mass Transf 128:1206–1217
5. Li ZH, Zhang H, Chen HP (2020) Application of transport membrane condenser for recovering water in a coal-fired power plant: a pilot study. J Clean Prod 261:121229

Chapter 7
Conclusion

This book focuses on the membrane absorption technology. To solve the problem that the membrane material is easy to be wetted, ceramic membranes with the hydrophobic property is prepared. It is used for the CO_2 capture in membrane contactors, in order to provide new ideas for the CO_2 capture technology in thermal power plants and other industries, and promote the clean and efficient development. Through the method of combining experiment and theory analysis, the membrane contactor is taken as the research object. The preparation of ceramic membrane materials, the mechanism of the gas mass transfer in the porous media and the mechanism of the CO_2 capture are studied in detail.

First, the influence rule of operating parameters on the CO_2 capture performance of the hydrophilic Al_2O_3 ceramic membrane is revealed, and a new technology of the CO_2 capture in the flue gas of thermal power plants by the ceramic composite membrane absorption is proposed.

Second, the membrane forming mechanism of the hydrophobic ceramic membrane is studied, and the preparation technology of the hydrophobic ceramic membrane applied in the CO_2 capture scene is proposed. The CO_2 capture performance of hydrophobic ceramic membranes is experimentally studied, and the mass transfer characteristics of CO_2 in the hydrophobic ceramic membrane are analyzed.

Third, the main factors affecting the hydrophobic properties of the membrane are analyzed, and the preparation technology of the superhydrophobic ceramic membrane for the CO_2 capture in the flue gas environment of power plants is put forward.

Finally, the interaction between the fly ash based ceramic membrane and the modified material is analyzed. Moreover, a low-cost superhydrophobic fly ash based ceramic composite membrane preparation technology is proposed, which can achieve the high efficiency and the low cost CO_2 capture.

Z. Li et al., *Hydrophobic Ceramic Membranes for CO_2 Capture*, SpringerBriefs in Energy, https://doi.org/10.1007/978-3-031-77678-6_7

Index